BASICS OF GEOMORPHOLOGY

About the Authors

Madhujya Mili, M.A. M.Phil., Assistant Professor, Department of Geography. Jagannath Barooah College (Autonomous), Jorhat, Assam.

Bhupendra Nath Goswami, M.Sc., PhD., Former Head of the Department of Geography, Mariani College, Jorhat, Assam.

Uttam Mili, M.A., Assistant Professor, Department of Geography, Hailakandi College, Hailakandi, Assam.

BASICS OF GEOMORPHOLOGY

Madhujya Mili
B.N. Goswami
Uttam Mili

CONCEPT PUBLISHING COMPANY PVT. LTD.
NEW DELHI - 110 059

ISBN: 978-93-5594-853-3 (HB)
ISBN: 978-93-5594-848-9 (PB)

First Published 2024

Published and Printed by

Concept
Since 1974

Concept Publishing Company Pvt. Ltd.
A/15-16, Commercial Block, Mohan Garden
New Delhi-110059 (India)
T: +91-11-41101460
E: publishing@conceptpub.com, *W*: www.conceptpub.com

Cataloging in Publication--Courtesy: D.K. Agencies (P) Ltd. <docinfo@dkagencies.com>

Mili, Madhujya, author.
Basics of geomorphology / Madhujya Mili, B.N. Goswami, Uttam Mili.
pages ; cm
Includes bibliographical references and index.
ISBN 9789355948533 (HB)
ISBN 9789355948489 (PB)

1. Geomorphology. I. Goswami, B. N. (Bhupendra Nath), author. II. Mili, Uttam, author. III. Title.

LCC GB401.5.M55 2024 | DDC 551.41 23

Preface

Geomorphology is more often regarded as the opposite foundation of the whole geographic discipline and it is at the same time certain that in absence of a proper distinct knowledge on geomorphology no other branches of geography can well be explained and make it feracious. Geomorphology is simply known as the study of landform those exist over the earth surface. A faire feracious concept on geomorphology will certainly enhance the opposite concept of geographic events and process and it is also certain at the same time that in absence of geographic concept the entire geographic analysis relating to landform will enfeeble the whole attempt to provide a scruples and objective based analysis on other geographic phenomena. Climate, economy, environment, culture, etc. of any region of the earth can be explicate with an edacious knowledge in geomorphology of the concerned region. At this backdrop, this book on geomorphology is designed as pertinent for the new generation learners. The entire text incorporated here in this book is made quite unambiguous in order to make it conducive and pertinent for the learners. Explication of all the contents incorporated here are arranged systematically to enamour the upcoming learners.

Generally, geomorphology is regarded as the science of landforms. Geomorphology appears as one discipline as a result of dissatisfaction with the term physiography. It usually includes climatology, meteorology, oceanography and parts of mathematical geography which in turn, impose a restriction in the landform study and in consequence, a proper scientific study on the development of landforms has come about as a result of dissatisfaction with the term physiography and in consequence landform study culminated as one basic part of geomorphology. The major contribution in the field of geomorphic studies is James Hutton, Playfair, Charles Lyell, W.M. Davis, Andrew Ramsey, G.K. Gilbert, Walter Penck etc.

Some of the significant trends of the past few decades are (1) a tendency for geomorphology to become more strictly geological than geographical as a result of (a) an increasing application to geomorphic studies of other phases of geology, e.g. mineralogy in the study of weathering, stratigraphic methods in paleogeomorphology and palaeontologic techniques in the study of glacial deposits, and (b) a decline of interest among geographers in physical geography as they give increasing emphasis to human geography; (2) the development of regional geomorphology, which attempts to divide the continents into areas of similar geomorphic features and history; (3) an increasing recognition of the practical applications of geomorphic principles to such fields as ground water geology, soil science and engineering geology, and (4) the down of the quantitative and experimental stage with attempts to apply the laws of hydrodynamics to a better understanding of the geomorphic processes.

Landforms those exist over the earth surface are basically products of contradiction between two opposites. The landform features are subject to undergone changes as a result of continuous interaction with the agents of denudation. Varying resistance of landform features and effectiveness of denudational agencies, in turn, give birth to varying landform features over the earth surface depending upon the nature and effectiveness of denudational agents works differentially in different parts of the earth surface. This logic is probably incorporated for the first time in geomorphology which in turn will make easier to understand the geomorphic activities over the earth surface.

The co-author of this book Sri Madhujya Mili, Assistant Professor, J.B. College (Autonomous) and Sri Uttam Mili, Assistant Professor, Mariani College has contributed a lot to make this book a standard one for the benefit of the students. The entire arrangement of the text systematized by these two co-authors from their experience they earned from their teaching experience. It will certainly be conducive one if this book become feracious for the learners of geomorphology and certainly it will be engrossing for us which in turn will rise the benefit of the learners. We are grateful to Shri Ashok Kumar Mittal, Concept Publishing Co. Pvt. Ltd., New Delhi for his unbidden co-operation and unbidden consent for publication of this book. We are also grateful to our precursor S.K. Borthakur for his inspiration in preparing a series of indispensable books for the students of geography.

B.N. Goswami

Contents

1

Geomorphology

Definition, Nature, Scope and Recent Trend

Definition

A systematic study on landforms is very simply known as geomorphology. Generally, it is considered as the science of landforms. Geomorphology is also more often defined as the systematic study of physical features of the earth's surface and their relation to the interest geological structures. Modern geomorphologists are however interested to give a wider view of geomorphology incorporating the features found in the ocean floor, which is ignored in the traditional definitions. Accordingly, Geomorphology is defined as a scientific study of the origin and evolution of topographic and bathymetric features (features of ocean floor) originated as a result of physical, chemical and biological processes operating at or near the earth's surface. The term geomorphology is, in fact, a product of three Greek root words—" Geo, morphe, and logos" which mean "Earth, shape and size, and study or discourse" respectively. Geomorphology, therefore, also is defined as—"geomorphology is the study of earth's physical features and the processes in which those features are formed.

Geomorphology originally was studying as a branch of physiography, which considerably includes part of climatology, meteorology, oceanography and mathematical geography. As a result of dissatisfaction on the term, physiography to study the whole mechanism of origin and evolution of landforms on the earth's surface, a new term 'geomorphology' was derived to study the origin and evolution of landforms of the earth's surface. The term 'geomorphology' became very much conducive to include all relating to the origin and evolution of landform features over the earth's surface and ocean floor.

Nature of Geomorphology

Nature of geomorphology is quite comprehensively attached with the root word itself, i.e. geo, morphe and logos. Geomorphology literally is the systematic study of the form or shape of the earth, but it primarily deals with the topographical features of the earth's surface and sea floor. Geomorphology, as some scholars prefer, is not merely a study of landforms and processes alone but also simultaneously emphasized on the importance of the interactions at the contact surface between the terrestrial part of the lithosphere and the liquid and gaseous envelopes which surround it and the ever changing inter-adjustments between different forces of nature.

Nature of geomorphology, however like other sciences can be illustrated with few rational aspects like—

(i) Description and interpretation,
(ii) Process and form,
(iii) Artistic and scientific elements,
(iv) Qualitative and quantitative aspects,
(v) Dialectics in Geomorphology, and
(vi) Laboratory and fieldwork.

(i) Description and interpretation

Observation is the primary task of all scientific enquiries. Descriptions are basically three types. These are observational, classificatory or explanatory and descriptive. German geomorphology traditionally prefers the importance of observations and was interested to provide a detailed description on the form and shape of various relief forms. Landforms having sufficient symmetry and distinctiveness of structure are classified as one specific type. On the other hand, a genetic description attempts to describe not only the form alone but also provides an interpretation on its origin, e.g. flood plain, pediment, sinkhole, etc. But increasingly there has been a trend to avoid lengthy descriptions with the development of specialized mapping techniques, As such another growing tendency is observed in the ways to separate description more interpretative. A. Cailleux and J. Tricart (1956) showed their interest in the use of scale of varying order with an aim to make geomorphology more meaningful in its scientific method of investigation and interpretation. Each and every details of landforms character can well be interpreted using mapping techniques. Use of suitable scale in maps will make geomorphological description more comprehensive and useful in the field of application. Ranking techniques, more often are suggested as an added advantage to make geomorphologic description more reliable and scientific. R.E. Horton was the pioneer in this field of ranking techniques in which run off prediction and channel behaviour can effectively be illustrated to enhance the descriptive nature of geomorphology and as such the rational interpretation.

Of course, there are many difficulties in devising precise, objective and unambiguous descriptions of the complexities of landform. Despite the time involved in such attempts, the results seemed to be a limited success. These factors ultimately help to explain the causes behind the reluctance to abandon the concise genetic descriptions in geomorphology. On the other hand, the incorporation of new informations and theories into existing interpretations give a stable frame of geomorphological description and thereby enhance the acceptability in the rising horizons of scientific studies. A major problem however is seen in the course of explanations of landform where intrinsic incompleteness of the geomorphological and geological evidences usually stands as a major constraint in the way of deriving a complete conclusion. Some geomorphologists advocated the use of quantitative methods to solve the problems faced with explanation. In geomorphological literature uncertainty, disagreement, and debate about the origin of specific landforms is common as the geomorphic processes are continuous and never ending components in the evolution of landform processes. The most in geomorphological research inspire his successors, Davis unequivocally announced—"we have learned to give greater praise to those who, on a good foundation of visible facts, employ a well trained constructive imagination in building ingenious and successful theories, which shall bring to sight the invisible facts."

(ii) Process and form

"Landform is a function of structure, process and stage" is considered as the basic corollary in geomorphological studies. Form is the ultimate product of process. A landform attained at its final state or ultimate state as a result of the processes worked upon it with varying intensity subject to the prevailing other natural condition such as climatic condition, soil quality, vegetative cover and topographic character. However, the most significant is the growing appreciation of uncertainty of the nature of most processes and interaction with landforms. The complex mechanism of ice movement, and sub-glacial processes are not yet completely came into knowledge. Such are the aspects of fluvial dynamics and relationship between the fluviatile processes and the reworking of depositional landforms.

The increasing interest in process studies developed as a result of complexities in the development of various landforms under varying conditions. Interpretation on landform development is integrated with geomorphological processes, which are seemed to be difficult to correlate and generalize them. There are numerous problems in studying and evaluating the consequences of process operating over certain landform. As a result, a new concept in geomorphological studies developed in recent time in form of dynamic geomorphology which in a general way relates to the interactions of form, process and soil. It is universally accepted that landform is the prime interest in geomorphology and also is the basic subject matter of geomorphology subject to a concrete interpretation thereupon. The systematic study of a distinctive form involves the study of processes as well. In principle, the geomorphologists attempt to infer processes from the forms and to study the processes associated with the development of various landform features. It is however, happened more often that no specific processes are alone involved in the completion it an ultimate landform. By form is meant the physical dimensions of a particular landform (shape and size). Two broad type of landforms are recognized, i.e. (i) Initial landforms (landform produced as a result of internal earth processes like volcanism, diastrophism etc.), and (ii) sequential landform (landform produced in consequences of external earth processes like erosion, deflation, weathering, mass wasting etc). However, the basic concern of geomorphology is to find the answer to three major questions about the earth's landform; the questions are:

1. How can these features and processes be described?
2. How can they form and change through time?
3. What processes are responsible for the resultant form and how do these processes work?

(iii) Artistic and scientific components

The prime need to understand geomorphology is a clear appreciation of nature the familiarity with the countryside, coasts, hills, rivers, mountain which develops interest in landform studies. Clear diagrammatic representation of landforms and associated features makes things easier to understand. Literary skill is also important in geomorphology in order to compile the growing range of qualitative descriptions. In developing new interpretations, skill is required in defining new terms with precision, clarity and with due merit to the previously used terms. In translating works on geomorphology from foreign languages, linguistic ability is necessary. It is moreover, an established fact that a clear and artistically well designed diagram to able to uphold the content of analysis of phenomena irrespective of any discipline. Eminent geomorphologist W.M. Davis emphasized on the quality of analysis by means of the qualitative presentation to make it comprehensive and simple.

During last recent decades it is seen that the scientific aspects of geomorphology has not neglected the need of artistic floor in geomorphic analysis. Along with artistic alienation geomorphologists must acquire sufficient knowledge on principles of hydraulics, meteorology, oceanography, botany, physics, chemistry etc. Geology is necessarily an integral part for geomorphic studies. In scientific method, the procedure from firsthand observation to interpretation is more important than discussion. Many more scientific techniques are borrowed from neighbouring science namely, soil science, hydrology, meteorology, physics, chemistry etc. for authentic geomorphic analysis. It is also believed that geomorphologist generalizations should be derived using principles of physics or chemistry. Increasing application of principles of certain science to the interpretation of landform necessarily opens a door for a fourth aspect in the study of landform which can be called scientific aspect. However, geomorphology is to some extent different from universal laws as landform study differs from region to region following their differential physical condition and the processes involved therein.

(iv) Qualitative and Quantitative aspects

The value of qualitative analysis behind the collection of data is one of the rudimentary aspects of quantification. Qualitative reasoning complemented by quantitative aspects of particular geomorphological studies always gains high acceptability and then becomes more authentic. But problem lies with the scale of measurement and the techniques employed in order to derive a common statement of universal characteristic. In the case of nature, the relationship between two phenomena is rarely that of a straight line. It is quite problem oriented that varying geomorphological process works differently following the variation of physical condition which practically generates problem in the path of generalization of form and processes. Although W.M. Davis and his followers emphasized more on the qualitative aspects to analyze the complex form and processes of geomorphology but it fails to fulfil the need of generalization and to derive the common principle on the evolution of landform. Qualitative and quantitative approaches characteristically are reciprocally interrelated. Quantitative measurements are considered as catalyst to qualitative analysis of geomorphic science as it enhances the scientific base of analysis of different aspects of geomorphology.

The importance of qualitative reasoning is attached with the collection of required data which is essentially a prime aspect of quantification. Recent development in technological know-how, however, reduces all the difficulties in the measurement processes and forms of geomorphic features. Unlike the other sciences namely physics, chemistry etc. geomorphology is not bound to any specific laws. In that sense, the law of drainage network derived by R.E. Horton is simply a useful generalization. Drainage network varies according to the varying nature of landforms. Making accurate measurements of all geomorphic feature and processes are quite difficult because of its extreme variability in size, variability in climate and variability in processes etc.

The measurement of the degree of association between two geomorphic variables is a difficult task. The relationship between two geomorphological phenomena in nature is seldom a straight line. Frequent change in the nature also imposes change in the geomorphic processes and the resultant features. In many cases, geomorphic relationships are characteristically curvilinear rather than straight. As such derivation of an equation among interrelated variables is important to enhance the authenticity of geomorphic studies. Under lots of limitation and uncertainty in establishing correlation and derivation of a genuine appropriate equation in the geomorphological studies, the quantitative aspect is felt necessary to augment the quality and acceptability of any form of qualitative analysis of all geomorphic forms and processes.

(v) Dialectics in geomorphology

Dialectical logic is primarily concerned with the interaction of opposing forces. "Contradiction between two opposite forces gives birth to a new one", following this law almost all the geomorphic features are the product of two opposing forces namely the endogenetic and exogenetic forces. The features came into existence over the surface of the earth as a result of endogenetic forces have to face the opposite forces acting upon from outside in form of exogenetic forces as it exposed over the surface. Contemplation on the forces acting actively over the surface and also in the interior of the earth would comprehensively elaborate the nature of all geomorphic forces and the landform developed in consequence. Endogenetic forces acts on the earth from below while the exogenetic forces acts in a total negative dimension with an intention to level up the surface. Running water as one most effective agent of soil erosion works against all the resistance imposed by the surface over which it flows. The contradiction between the running water and the resistance created by the varying nature of the surface condition in consequence changes the whole structure of the landform and in turn gives birth to a new feature which is totally new in its structure and form. Hills and mountains which are virtually product of endogenetic forces are converted into a level surface in form of plain by the continuous works of denudation process, inclusive to exogenetic forces.

Qualitative change from quantity is the second important logic of dialectives. While looking into the various geomorphic processes it is quite convincing that increasing intensity of denudation and deposition ensured changes in the quality of the resultant landform which is true everywhere and any time over the surface of the earth. Increasing amount with rising intensity of geomorphic agents brought a severe change in the quality of the landform over which they are active and effective. Increasing quantity of either sea waves or winds or running water certainly brought qualitative change in the landform as such reverse may be the case for decreasing intensity of the processes.

Voluminous glacial flow, as such, cause severe changes in the area along which it move in comparison to the effect of smaller noval size glaciers. Geomorphic processes are found in varying degree and intensity in different parts of the earth surface sufficient to the natural environmental condition. Increasing number of ocean waves also cause larger and more impressive effect on the shore areas in comparison to the normal periodic waves. It is one established fact that quantity determines the quality specifically in case of all geomorphic processes actively working over the earth surface with varying degree and intensity.

Negation of negation is the third law of dialectical logic. The geomorphic processes consistently working upon a specific landform causes changes of the landform negating the earlier shape, size and structure under a normal condition. Hills and mountains are concreted into broad and flat alluvial plain, a featureless plain desert may be filled with loess and sand dome, a narrow steep sided coast may be converted into an extensive coastal belt etc. are few examples of negation process. On the basis of their logic almost all geomorphic processes can be explained and elaborated with all judicious rational manners. In explaining all their geomorphic events under the light of dialectical logic it will be quite necessary and a compulsion to incorporate the techniques of measurement, fieldwork, mapping etc. Measurement, fieldwork, mapping etc. are integral to the dialectical explanation of all geomorphic events.

(vi) Laboratory and Fieldwork

Primarily, geomorphology has close association with the field observation, measurement and mapping of various landform features. No accurate and reality based rational study of various landform features under circumstances of continuous change is possible without having extensive field observation and laboratory work. Geomorphologist at the primary stage of geomorphological studies were characteristically explorers or the field geologist preparing maps of landform features and the processes on the basis of mere reconnaissance survey as there were no sufficient required tools and techniques. However, the most fundamental geomorphological experience can be gained only through looking the realities of landform. Because of the observing landform will provide a direct experience of wide range and enhance the accuracy in the process of generalization. Field observation virtually enhances the chances of the discovery of crucial evidence either as a distinctive assemblage of forms or an unusually complete stratigraphical section. Since 1945, the growing emphasis in some aspects of landform study has tended to pass from exploration to productive systematic research. Any form of secondary sources for geomorphic interpretation and analysis of landform is not at all authentic as the whole of the landform is subject to undergone continuous changes as time passes. Day to day dynamism of landform development can well be studied only under continuous strict and direct observation. At this backdrop, extensive systematic fieldwork is the most pertinent and opposite means to analyze and interpret the inherent geomorphic nature of the varying landform over the earth surface. Modern photographic technology or the remote sensing imageries can better be used as supplementary to the field observation but in no way can be used as an effective alternative to the direct field observation to acquire an accurate idealization on the dynamic nature of geomorphic processes. Denudation, transportation deposition etc. by air, water, glacial and their molecular constituent along with the heating and cooling by sun shine can effectively be depicted by direct field observation only and in no way such inherent variation in all the processes can be traced with total accuracy which eventually enhance the scope of new discovery on landform studies.

Data and informations pertaining to sediment characteristics, flow dynamics, depositional behaviour and stratigraphy etc. are needed to bring under laboratory works to analyze co-relation, classification, generalization, experimentation, prediction of possible form and structure, direction of change, finalization of findings etc. in order to establish a scientific base of field observations. Extensive field observation is fundamental to enhance the acceptability of geomorphic principles so established. More generalization of geomorphic form and processes will be quite misleading under varying climatic, stratigraphic and structural differences from region to region.

Rising acceptance and importance of scientific findings in the laboratory experiments and also the authenticity make geomorphology a subject of scientific orientation. With the increasing emphasis on the importance of field measurements and same collection for laboratory experiments practically enhance the scope of fieldwork. C.A.M. King (1959) advocated that engineering scale models do simulate with reasonable accuracy in the changes of natural features and thus interest in scale models has grown among the geomorphologist and geologist. However, it is fact that with the increasing tendency for field measurement and sample collection for laboratory analysis, the importance of field study still remains relevant in geomorphic studies. Field observation and laboratory work is reciprocal to each other to have a distinctive study of geomorphology.

Scope of Geomorphology

As a whole the geomorphic ideas are solely concerned with all the surfacial characteristics of the earth, the interacting zone between the living creatures and the natural environment integrated with the surfacial form of the earth. Geomorphology as a study of the landform features of the earth is closely associated with innumerable aspects relating to the origin, evolution and the ultimate resultant structure of a specific landform features. It is, therefore, necessary to have an interdisciplinary approach to fabricate a sound positive knowledge base on the varying landform features starting from its origin to the present day form of the landforms. The description and analysis of landform is a science when it is studied in a systematic disciplined fashion. Geomorphology is characteristically an interdisciplinary subject as it combines various aspects of geology, climatology, physics, chemistry, sociology, mathematics etc. to provide concrete ideas on landform development starting from its origin to the ultimate stage of evolution. The scope of geomorphology is, therefore, very extensive and required a wide range of knowledge of various disciplines.

Geological content in geomorphology is quite extensive and therefore, geomorphology only can thoroughly be studied with an in-depth knowledge in geology. Geomorphology brings a study of the landform features of the earth concerned with the landform structural history of all the landforms to provide a rational and orderly analysis and exploration on the evolution of the landform features. The origin of the landform is solely the concern of geological processes and to study the evolutional dynamics of the landform, the originality in no way can be ignored. Moreover, landform almost invariably imposes some degree of resistances to emotion a weathering of the underlying rocks. Since all the initiate landform are the resultant product of the forces originated within the earth, i.e. the endogenetic forces, therefore, knowledge of geology as the genesis and development of all types of landform is a prerequisite to study the geomorphic nature of the landform development.

Landforms almost invariably impose some degree of differing resistance to erosion and weathering processes and the degree of resistance depend on the nature and composition of the rocks. For instance, in pre-Triassic rocks of Connecticut quartz rich rocks stand relatively high, whereas carbonate rocks, foliated rocks rich in micas and poor in quartz and paleoglacial rich rocks generally underlie lower areas (A.F. Pitty, p. 91). A clear and distinctive knowledge on the nature of rocks and structure of the surface is necessary to have an extensive and accurate concept on the works of the geomorphic processes which are more responsible for the development of varying landform features over the earth surface. Petrology, Mineralogy, Crystllography, Stratigraphy etc. various branches of geology act as supplement in acquiring perfect ideas on the whole process of landform development.

Most of the landform units of the earth surface owe a great deal to uplift of land over extensive areas in the geologic post. Variations in surfacial form like, hills, mountains, plateau etc. are final consequence of tectonic activity of the earth. Hardness of rocks, stratigraphic arrangement of rock strata, composition of rocks, nature of slopes etc. are basic components of all geomorphic analysis. It is now everywhere common that application to geomorphic studies of other phases of geology becomes a compulsion without which no geomorphic study can be completed, e.g. mineralogy in the study of weathering, stratigraphic method in the study of paleo-geomorphology, paleontologic techniques in the study of glacial landform etc. Geomorphology in the past was studied as a branch of geology and especially in U.S.A. still today geomorphology is treated as a major branch of geology to study all the landform features.

Geology-Climatology

Solar radiation is the sole driving force of all the chemical, biological, and mechanical forces of sub-aerial weathering processes. Intensity of solar radiation decreases pole ward and thereby leads to excessive cooling which is counteract by atmospheric and oceanic circulation and prevent temperatures at the poles and equator from becoming progressively more extreme. Water, an indispensible component for weathering, erosion and removal of weathering products occurs naturally on the earth surface. Climate determines the availability of water and its volume and period of flowing stream and rivers, over soil surface and down slopes. Climate varies from region to region and thus determines the nature and intensity of geomorphic processes. Over the earth surface heavy downpour in the tropics and monsoon region enforced, intensive channel flow and surface run-off and leads to extensive soil erosion in those areas. Mechanical and chemical weathering of rocks is also closely associated with the prevailing nature of climate.

Exogenetic forces are virtually the product of climate which work upon the landform as agents of all degradational and aggradational processes against the endogenetic forces. Temperature, pressure, humidity, precipitation, storm, cyclone etc. all are climatic phenomena by their nature and origin. Without having a perceptible and comprehensive study on the nature, form, intensity etc. of all these phenomena, it is quite impossible to provide a perfect and accurate study on the changing nature of landforms. Variation of climate from place to place and their cause effect processes are closely associated with the variation in landform features over the earth surface. Landform developed in areas of heavy rainfall are characteristically different from those developed in the arid region and similarly landform developed in the glaciated areas are different from other landform in shape, size and structure developed in other non-glaciated areas. The shape size and structures of landform differs as the geomorphic processes, namely erosion, weathering, deposition etc. differs following the prevailing climatic condition in varying parts of the world. Any form of procrastination on climatic variability prevails from region to region will certainly detract the attempts made to study the dynamism of landform development and will obviously be resulted in a total deterioration of the facts attached with all kind of geomorphic processes. Acceptable scientific explanation and analysis on landform development should, therefore, be studied with second concern of all climatic components. Degree and intensity of erosion and weathering depends on the climatic factors and elements which naturally varies from region to region. No accurate geomorphic studies on the process of landform development can be accomplished in an appropriate way without keeping the climatic condition aside. Climate controls processes and processes control form, therefore, form is a product of climate (Khullar, 2012).

Geomorphology and Other Sciences

Landforms are primarily the resultant features produced as a result of earth's structure, its gravitational force and its magnetic properties structure of the earth is the concern of geology while the gravitational force and the magnetic properties of the earth is studied in physics. All landform are composed of different rocks which are different in their chemical composition and molecular content. As such strength and weakness of rocks, their stratification, mineral composition etc. are some of the important components of landform development. This is one of the basic concerns of geomorphology as it is associated with all the processes of landform development.

Along with these, the human acting in form of agriculture, industry, hydro-power project, trade, tourism, urbanization etc. are major subject matter of geography. Since there activities are relating to the nature of the surface and subject to changes brought in the surfacial form, therefore, geomorphologists in no way can procrastinate their activities in order to provide an accurate geomorphic analysis on the landform development as a whole. All these activities are somehow influenced by geomorphological features such as mountains, plateaus, plains, deserts, lakes, river valleys, sea coasts etc. Geomorphology is, therefore, an integral part of geomorphical studies.

Geomorphological features are naturally formed of various rocks. Critical analyses of the rock types are necessary for a comprehensive study of landform. An accurate chemical analysis of rock types is therefore, mandatory to augment the level of accuracy in the analysis of landform development. Chemical content of rocks response differently to the processes of landform development and therefore, geochemistry is also integrated with all forms of geomorphic studies, composition of soil and their chemical content is important for qualitative geomorphic studies of various landforms. Any form of ignorance to this aspect will somehow cumbersome the whole geomorphic approach in studying the various landform features over the earth surface. Chemical and mineral composition varies with the varying kinds of soil, details of which can be supplemented through a careful chemical analysis of the soil quality. Chemistry with this essence is closely attached with all kinds of landform studies.

Dynamism of landform development is one basic component of geomorphological studies. Evolution of all landforms can be explained well following the dynamic behaviour of land, water, lava, magma, wind, glacier which are the basic components of the evolutional processes of all landform visible over the earth's surface. Knowledge on physics only can provide a scientific analysis of landform development from Precambrian to the present Holocene period. Similarly, bio-sciences are also important to comprehensive geomorphic studies. Bio-science is primarily concerned with the study of vegetation, human being and all other living animals on the surface of the earth. All these are related with the process of landform development in an area. Dense vegetative cover in an area works as a protective component of soil erosion and weathering. Overgrazing by animals destroys the vegetative cover and below prone to massive erosion by water, wind and glacier. Man on the other hand is a great modifier of landforms. He brought a massive change in the landscape by expansion of agriculture, settlements, (both rural and urban), mining, communication etc. Thus, knowledge of bio-science becomes essential for all kinds of geomorphic analysis.

The scope of geomorphology is practically much more extensive and therefore, an integrated approach for a rational geomorphic analysis of landform development becomes a compulsion. Analysis and explanation derived on the basis of such a multidisciplinary approach in consequence enhance the scientific base of geomorphology and also its acceptability in the relevant field of application.

Recent Trends in Geomorphology

Alike other sciences, geomorphology has made rapid progress during the 1960s in form of techniques, concepts and aims. The basic changes observed in the increasing input of quantitative methods, primarily of those depending on data processing by computer. Quantitative approach chiefly concentrates on fluvial morphology, morphology of arid and semi-arid zones, the humid tropics and glaciated areas. R.E. Horton's work in 1945 opened up a new dimension into the study of significance of drainage topology. Quantitative techniques he used to analyze the drainage network got widespread acceptance which however in turn replace the simple descriptive method earlier used to analyze the geomorphological events. Extensive and meaningful use of quantitative techniques provides a solid scientific foundation of geomorphology. The mega and meso-scale used earlier in the study of landform is replaced by micro-scale. It enhances the field measurement and instrumentation to procure authentic information on various geomorphic processes. It, in turn, helps geomorphologist to develop functional theory of landscape development. Introduction of system theory for a scientific explanation of landform development helps to formulate various geomorphic models. Model building on geomorphic processes and landform development augments the understanding of all complex geomorphic evolution of landscape in a meaningful way. The models developed in geomorphology can be grouped as natural analogue system, physical system, general system etc. Moreover, process response model is now more popular and become focal theme of process geomorphology. Geomorphic models can very distinctively explain the co-relation between the processes and the evolution of a landform which now appear as one common trend among the geomorphologist.

Environmental geomorphology appears in the field of geomorphology as one important aspect with the growing environmental crisis. In order to keep pace between the development and environment as most of the environment problems are characteristically development induced one, therefore, an appropriate analysis on the geomorphic processes becomes a compulsion to continue the developmental programmes. Geomorphic knowledge for the removal of environmental problems becomes a necessity to understand the interactions of economic and technological components with geomorphological process and the existing natural system. Impending environmental crisis of today becomes a serious concern of all authority engaged in developmental works and it is now a compulsion for those authority to make an assessment of the environmental impact prior to pursue development projects. For example, environmental impact assessment is now mandatory prior to construct hydro power project where air, water, soil, ground condition, vegetative cover etc., the major components of environment are carefully studies on the best of the character of the landform.

Another significant trend of geomorphology is to be more geological than geographical. It is indubitably a resultant product of increasing application of geological content in the study of geomorphology. Various branches of geology become essential to provide an accurate and appropriate study on landform development, for example, Mineralogy in the study of weathering, stratigraphic methods in palaeogeomorphology, paleontologic techniques in the study of glacial deposits etc. In view of many geomorphologist's knowledge on geology is a prerequisite to understand well the evolutional processes of the landform and therefore it become trend in geomorphic studies. The extended use of remote sensing techniques helps the geomorphologist to cover up more areas which are practically inaccessible to conduct fieldwork. Moreover, application of remote sensing techniques ameliorate the geomorpholgical studies in relation to landform development by way of providing all necessary up to date information through landsat imageries and photographs. It enhances the laboratory work for the geomorphologist negating the hazardous, laborious time taken fieldwork. Development in computer software technology to analyze the imageries and photographs procured of varying magnitude from satellite enlarge the horizon of geomorphology with all expected accuracy. Introduction of high resolution on board camera in the satellite system so designed and developed that all requirements needed for landform studies become effective and available.

Areas covered by impenetrable jungles, unexplored deep sea floor can now been studied well with the help of the imageries procured from satellite. The general trend of geomorphology presently becomes more technology oriented over the earlier descriptive mode of theoretical aspects.

2
Interior of the Earth

The knowledge of the interior of the earth is very limited. Through direct means by mining and drilling, we have been able to penetrate and examine directly an extremely small part of that distance. The lure of gold has taken people (specifically miners) to a depth of 3.5 kilometers in South Africa, while drilling for oil and gas has taken our machinery to a depth of about 12 km. These explorations have just barely scratched the earth's surface. Geologists are continually working to understand the interior of the earth better. Extending scientific knowledge about the structures, compositions and processes operating within earth helps us to learn more about such lithospheric phenomena as earthquakes, volcanic eruptions, the formation of rocks and mineral deposits and the origins of continents. It can even help us to learn more about the origin of the planet itself.

From the theories relating to the creation of the earth, it is clear that the structure of the earth has evolved through burning vapour and various changes over the time. Scientists have almost agreed that the earth once existed as a burning vapour, and they agreed that due to radiation of heat it gradually cools and concentrate the body. The burning vapour contained different gases and metals such as water vapour, iron, nickel etc. The temperature of all those different gases, their radiation level and density were varying to each other. As a result, lighter gases such as water vapour etc. were easily converted into liquid than the heavy metals. Again all those fluids changed in their relative importance, and relatively heavy metals such as iron, nickel etc. began to accumulate towards the centre of the earth and lighter materials on the upper part respectively. This process was continuing for many years, also reduced the amount of heat in the interior of the earth and the surface liquid as well. Finally, the upper part of the materials cools and creates a solid outer layer of the earth.

Seismic waves and interior of the Earth

Although it is possible to directly or indirectly know about the interior rocks through volcanic lava flows by conducting various experiences on the outer surface of the earth, in fact it is possible to know in details the internal structure of the earth by analyzing the velocity of the vibration of seismic wave. Disturbances can be noticed when the waves are interrupting by opposites, because of their characteristics changes as they move from one rock to another with the help of seismograph. Based on the velocity, it is possible to know what kind of wave is passing through the rocks. These waves move at different speeds depending on the structure of the earth. Most of the earthquakes originate at a definite depth below the earth's surface. The depth may differ from a few meters to several kilometers. The place in the earth's crust, where the earthquake originates is called the **"focus"** of the earthquake. The point on the earth's surface directly above the focus is called **"epicenter".**

Classification of seismic waves

When the blocks slip suddenly at the time of the earthquake, intense vibrations called **"Seismic waves"** (from the Greek seismos, meaning **"shock"** or **"earthquake"**. Seismic waves are divided into three divisions according to their length and penetration. These are—

(a) Primary waves,
(b) Secondary waves, and
(c) Surface waves.

Primary and Secondary waves cannot penetrate through liquid materials as it can move through solid only. The velocity of these waves helps to gain an understanding of the nature of the rock because, the speed of the waves changes according to the nature of rocks. These waves are discussed below—

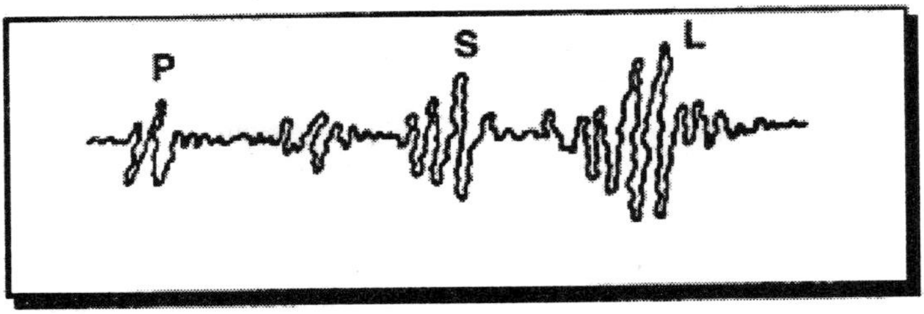

Source: Singh, Savindra: *Physical Geography* (2010), Prayag Pustak Bhawan, Allahabad, p. 48.

Fig. 1: Seismographic recording of P,S and Surface wave or L wave from a distant earthquake.

The cross-section shows the paths followed by the three types of waves.

(a) Compressional/Longitudinal/Primary waves or P wave

This is the fastest wave. Its speed is 5 to 8 kilometers per second. The frequency of these waves is very high and the wavelength is also very wide. P wave can travel through solid, gaseous and liquid materials. This means that the wave can travel directly by piercing the center of the earth. However, these waves can move faster through solid objects than liquid. Their speed varies according to the density of the medium through which they are travelling. Their speed increases to as much as 15 km. per second in the medium of very high density. The P wave is identical to the sound wave. The knowledge we gain about the structure of the earth's interior has been made possible mainly through P waves.

(b) Secondary wave or Transverse or Distortional wave, Shear wave or S-wave

The velocity of this wave is 3 to 5 kilometers. S waves travel only through solid rock at about half of the speed of P waves. The frequency of this wave is very high but its length is very low because these waves cannot move through liquids or gases. Consequently, **S** waves reach the recording station after the **P** waves have arrived. For that reason they are designated by the letter **'S'**. Again, the velocity of the wave gradually changes according to the density.

(c) Surface waves or L waves

Surface waves are confined to the earth's surface and outer layers because, like waves on the ocean, they need a free surface to ripple in order to exist. Their speed is slightly less than that of **S** waves. The length of the wave is maximum and frequency as well. For this reason, they are called **'L'** waves. These surface waves are governed by the formation of small angles with the rock. They affect only surface of the earth and vanish out at a smaller depth. These waves causes all the major damages to all the structures lying over the surface of the earth.

Seismology and Constitution of the Earth's Interior

The effect of seismic waves witnessed throughout human history. But not until the end of the nineteenth century, seismologists were not able to devise instruments to record such waves which enable us to locate earthquakes and determine the nature of faulting, and they provide our most important tool for probing the earth's deep interior. B.B. Golitsyn rightly remarked, "earthquakes are lamps illuminating the structure of the earth's interior."

If earth were made of a single material with constant properties from the surface to the center, **P** and **S** waves would travel along a straight line through the interior from the focus of an earthquake to a distant seismograph. But in reality, the waves do not follow a straight path, from which scientists infer that the earth is layered and is made up of many materials that conduct **P** and **S** waves at different velocities. The waves bend as they move from layer to layer and as a result their paths through the interior becomes curved.

Fig. 2 shows the curved paths of **P** waves as they travel from an earthquake focus to distant points where seismologists recorded on a seismograph. Follow the path of a wave that just misses the core and emerges at the surface at angular distance of 105°. Thereafter, follow the waves that strike the core. Interestingly, **P** waves bent downward when they enter into the core and are bent again when they leave.

Fig. 2 (a) shows the pattern of **P** wave paths through Earth's interior. Dashed line indicates the progress of wave through the interior at 2 minute intervals. The P wave shadow zone extends from 105° angular distance from the focus to 142°. **P** wave cannot reach the surface within this zone because of the way they are bent when they enter and leave the core. Fig. 2 (b) the **S** wave shadow zone extends from 105° to 180°. Although **S** waves strike the core, they cannot travel through a fluid zone and therefore never emerge beyond 105° from the focus.

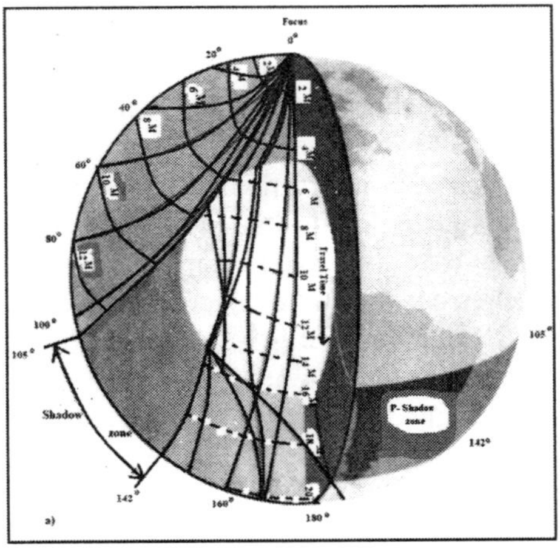

Fig. 2(a)

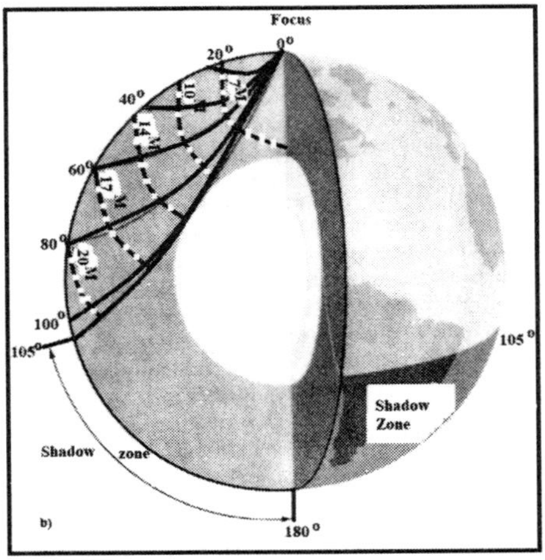

Fig. 2(b)

Source: Frank, Press & Raymond Siever: *Understanding Earth* (2001), New York: W.H. Freeman, p. 184.

Because of this bending at the mantle-core boundary, none of these waves emerges at the surface before 142° angular distance from the focus. Thus, no **P** wave reaches the surface between 105° and 142° as though the core had cast a shadow across this zone. This in fact is called the **"shadow zone".**

Discovery of shadow zone helps scientists to suspect that the earth has a core made of a material different from that of the overlying mantle. They could even conclude that the core is liquid, because decreasing velocity and direction of waves bent downward rather than upward when they enter into the core. It means that the waves travel more slowly in the core than in the mantle. **P** waves are known to move much more slow in liquids than in solids, as it was reasonable to guess that the existence of the shadow zone implies a molten core. This surmise can be verified by one other piece of evidence as the behaviour of **S** waves (Fig.2 b). Whenever **S** waves strike the core, they fail to emerge from the other side and are not seen at all beyond the 105° angular distance from the focus. Thus, scientist concluded that a molten core must have been blocking **S** waves from reaching these distances.

Waves Reflected in the Earth

Reflected wave implies a boundary between two materials, a significant way to find boundaries in Earth's interior is to find reflections when **P**

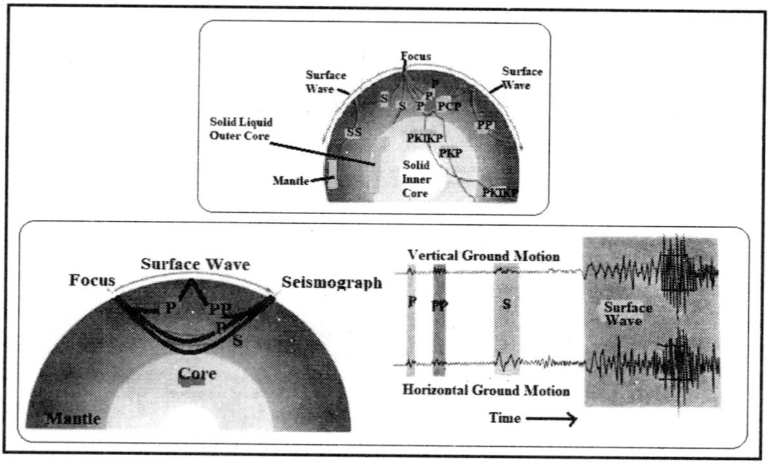

Source: Frank, Press & Raymond Siever: *Understanding Earth* (2001), New York: W.H. Freeman, p. 186.

Fig. 3: Seismograph recording of P, PP, S and surface waves from a distant earthquake.

and **S** waves bounce off the boundary between two layers, **PP** bounces from the Earth's core back to the surface. Because the velocity of a **P** wave is known, the round trip travel time for **PP** is used to determine the depth to the core, just as the time it takes an echo to return can be used to measure the distance to the wall on the other side of a valley. **P** waves can also bounce off the Earth's surface back into the Earth's, as shown by the PP wave in Fig.2. Similarly, **S** waves that reflect back into the Earth are called **SS**. The **P** waves that penetrate the outer core (PKP) or the inner core (PKIKP) are useful for exploring those regions. Fig.3 shows seismographs in which the **P, PP, S** and surface waves are indicated. They are differentiated from one another by their velocity and frequency. In general, an earthquake generates all the wave types shown in Fig. 3.

Classification of Earth's Interior

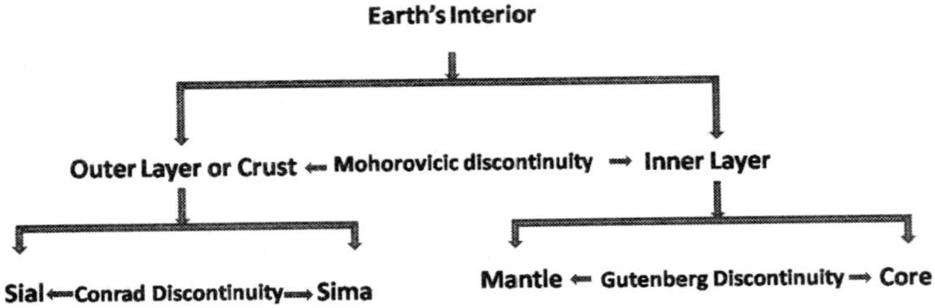

Fig. 4: Interior of the Earth.

1. Outer Layer of the Earth/Crust

It is a difficult task to collect information about the variation of the density of earth's elements and the composition of their various masses, but the depth or nature of this outer surface depends on the variation of the velocity of seismic waves. It is clear from the various soil layer of the upper part, consists of soil, stones, sands etc. that composed the outer layer of the earth. Depth of the outermost layer of the earth varies from place to place. This difference is reflected by the variations in the longitudinal velocity of the seismic wave. The interior part of the continent, the velocity is about 7 km/sec. But at the bottom of the ocean, the velocity increased to 8 km/sec. Such variations of seismic velocity indicate the density of the rock structure in the earth's interior. The existence of continental layer can be seen up to about 30-40 km.

depth from the surface. On the other hand, the depth in the bottom of the ocean is about 3-10 km.

This layer is located just below the sedimentary layer of the earth crust and generally composed of granite layer. This layer called **SIAL** because the first letter of silicon (Si) and Aluminum (AL) makes it. Silicates of aluminium, sodium and Potassium are found in this layer in large amount. The average density of this SIAL layer varies from 2.75 to 2.90 g/cm^3. The rocks of this layer are acidic, which is due to the great granitic mass occurring in the continental shields, and also on account of the great masses of sandstone in the sedimentary rocks which must have been derived primarily from granite.

There is another layer just below the **Sialic layer**, known as **SIMA**. It is made up of the first letters of **silicon** (Si) and **magnesium** (Ma), the two most abundant elements present in this layer. This layer is extending from 100 km to 2,900 km. The average density of this layer is varying from 2.90 to 4.75 g/cm^3. The silicates of magnesium, calcium and iron are most abundantly found in this layer. This layer is primarily composed of basalt that is the basic source of magma and lava.

Depending upon its thickness and composition, the crust can be recognized in three different types:

(i) Continental crust: Continental crust comprises the major landmasses on Earth that are exposed to the atmosphere. This layer is less dense (2.7 g/cm^3) than oceanic crust, with an average thickness of 32-40 km. It is also much thicker than oceanic crust. At places where continental crust extends to high elevation in mountain ranges it also descends to great depths below the surface. Continental crust contains more light colored rocks than oceanic crust does, and can be described as granitic in composition. The **L** waves travel through this layer with velocity of 1.5 to 5. 0 km/sec. The velocity of seismic waves increases with increasing depth (about 6 km/sec).

At the bottom of the continental crust there is a layer composed of basaltic elements. The density of this layer is 2.8 to 3.3 g/cm^3. The L waves move at the rate of 6.0 to 7.6 km/sec through 40 km thick layer. "Conrad Discontinuity" traced at depths of 10 to 30 km separates the granite and basaltic layers from each other.

(ii) Oceanic Crust: Ocean crust composed of heavy, dark-colored, iron-rich rocks that are also high in **silicon** (Si) and **magnesium** (Mg). Compared to continental crust, oceanic crust is quite thin because its density (3.0 g/cm^3) is greater than that of continental crust (2.7 g/cm^3). Forming the vast, deep ocean flows as well as lava flows on all of the continents, basaltic rocks are the most common rocks on Earth surface.

(iii) The Transitional Region: This Region is usually found on the peripheries of the major continents where there are marginal seas and islands located. This type of crust occupies intermediate place between the continental and oceanic crust.

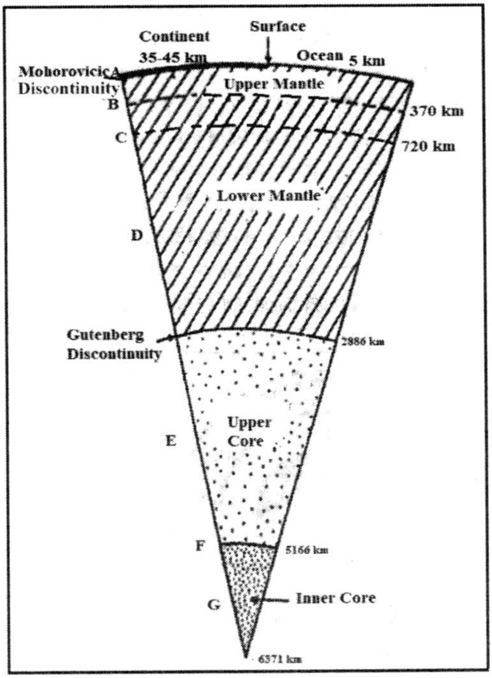

Source: Mitra, Kalpana, Sen, Jyotirmoy & Sengupta Purabi: *Physical Geography* (2014), Kalyani Publisher, Ludhiana, p. 25

Fig. 5: Structure of the Earth's interior.

2. The Interior Structure

As the seismic waves enter into the interior, the velocity also increases. It indicates that the density and pressure is increasing with increasing depth. According to the structure of the interior, it is divided into two divisions:

(i) Mantle, and
(ii) Core.

(i) Mantle

The thickness of the Mantle is approximately 2885 km. and representing nearly 85% of Earth's interior zones. Earthquake waves that pass through the mantle indicate that it is composed of solid rock materials, in contrast to the molten outer core lies beneath, it is also less dense than the core, with values ranging from 3.3-5.5 g/cm^3. Here, the average velocity of **P** waves increased from 6.5-7.0 km/sec in lower part of the earth's crust and it increases to 8.0- 8.2 km/sec to 4.5-4.7 km/sec. This interface between Mantle and the overlying crust is marked by a significant change of density, called a **discontinuity**, which is indicated by an abrupt increase in the velocity of seismic waves as they travel down through this internal boundary. Scientists called this zone the **"Mohorovicic discontinuity"**, or **"Moho"**. It was first established by the Yugoslavian seismologist "Mohorovicic" in 1909 and is known as "Mohorovicic discontinuity" after his name.

Based on structure, Mantle can be divided into two divisions—

(i) The upper mantle form the Mohorovicic Discontinuity to the depth of 1000 km, and
(ii) The lower mantle from 1000 km to the Wiechart-Gutenberg discontinuity.

(i) The upper mantle: The upper part of the mantle, with an average thickness of about 1000 km, is relatively cool, hard and strong. This contrasts sharply with the hotter, weaker material in the next lower mantle layers that flows particularly in response to applied stress. The outermost layer of this part has a chemical composition like the rest of the mantle, but it responds to applied stress more like the overlying Earth's layer, the crust. Together the uppermost mantle layer and the crust form a structural unit called as **lithosphere**. In recent decades, however, the term lithosphere has also been used in a separate, structural sense to refer to the more rigid outer shell of Earth, including the crust and the uppermost mantle layer. This layer is characterized by a marked increase in the velocity of seismic waves with increasing depth.

Extending down from the base of the lithosphere about 500 km is characterized by decreasing of velocity with depth. This layer is called **"Gutenburg's discontinuity**. This part of the Gutenburg layer is also called the **"Asthenosphere"** (from Greek asthenias, without strength or weak layer), a less dense layer of softened rocks. The material in the Asthenosphere can flow both vertically and horizontally, dragging segments of the overlying, rigid lithosphere along with it. Earth scientists now believe that the energy for tectonic forces comes from large scale forces that break and deform earth's crust. Movement in the asthenosphere, in turn, is produced by thermal convection currents which occur in the rest of the mantle below the asthenosphere and which are driven by heat from decaying radioactive materials in the planet's interior.

(ii) The lower Mantle: This layer is quite thick which is about 2000 km and consists of two layers (a) first layer extending from 1000 km to 2700 km, and (b) Second layer extending from 2700 km to 2885 km. Lower mantle is made of magma, under great pressure.

(ii) Core

Earth's innermost section, the core, contains one third of Earth's mass and has a radius of about 3360 km, which is larger than the planet Mars. Earth's core is under enormous pressure—several million times atmospheric pressure at sea level. The boundary of the core is just beneath the lower mantle, extending from 2885 to 6371 km. This layer is composed of **Nickel** (Ni) and **Ferrous** (Fe) and is known as **NIFE** (Ni+Fe). As it is composed of nickel and iron, it has very high density that reaches 13 gm/cm^3. The presence of iron is responsible for the Earth's magnetic properties.

The core is further divided into two sub-zones, e.g. **outer core** and **inner core**. The outer core extending from 2885 km to 5150 km. Rock matter at top of the outer core has density of about 10 gm/cm^3. Earth scientists proved that the outer core is in molten state (melted/liquid rock matter). The high density of both sections of Earth's core supports the notion that they are composed of iron and nickel.

Earth's inner core has a radius of about 960 km. The velocity of seismic waves travelling through the inner core shows that it is a solid with a very high material density of about 13 gm/cm^3. The material of the inner core is under higher pressure than the rock matter in the outer core. As a result of the great pressure, the material in the inner core remains solid despite its high temperature in the outer core are lower than in the inner core. But pressures are lower there as well, and this cause the outer core material to exist in the molten state. Internal temperature are estimated to be 6900°C at the very center of earth, decreasing to 4800° C at the top of the outer core.

Temperature condition of Earth's Interior

It is not possible to assume the temperature condition of the earth's interior by the living beings of the earth surface. In fact, under the influence of immense pressure on the surface of the earth, all those heated substances remain in a very stable state. Nevertheless, sometimes due to the decreasing pressure inside the earth, the heated material comes out to the surface of the earth. The events of volcanic eruptions prove this true.

Scientists almost agreed that the earth was a burning gaseous form in thorough radiation process, the outer layer of the earth's surface gradually getting cool but the interior is still very hot. The temperature of the earth's interior is not evenly distributed. Temperature rises steadily with increasing depth. Several experiments prove that although the temperature increases to 300°C per km, the rate of increase is not the same everywhere. Experiments have also shown that the rate of increase of the temperature up to the first 20 km. is almost same, but in the later case, the temperature-increasing trend is uneven from place to place

Table 1: Correlation of temperature rise with increase in depth

Depth (in km)	Temperature (°C)
20	550
40	800
60	1050
80	1200
100	1350
200	1700

Although it has long been assumed that, the formation of earth from heated gaseous state to the cooling of the surface but interior is still heated. Current geological experts prove that the large amount of radioactive material is present in the earth's interior. These substances always aid to increase the heat of the atmosphere by radiating heat.

It is not possible to have a solid object at such an immense heat in the interior of the earth. Therefore, different rocks and minerals in the interior are bound to get liquid state. But in reality such a situation is not seen in the interior of the earth due to the immense pressure of solid crust. For example, in normal conditions, the basalt melts at 11500°C, but if a pressure of 30 tonnes is applied per square centimeter, the temperature will increase to 14000°C. Therefore, the melting point of the interior material is increased only for the pressure inside the earth. Again, the temperature of the atmosphere also proved that the inner substances remained in a flexible and elastic state, not under melting or solid, under the influence of pressure and heat.

Pressure and Density of the Earth's Interior

As heat penetrates into the earth's interior, its pressure and density also increases. However, the rate of increasing density is not uniform. This increasing density depends on the effect of gravity and the increasing rate of temperature. Lot of experiences proved that the increase in the density of the object at the rate of increasing amount of pressure. It is found that the average density of the earth is 2.7 gm/cm^3 but in the 200 km depth, the density increases to 3.40 gm/cm^3. Again, in the 2900 km depth, the density is 5.5 gm/cm^3 and this density appears to be abnormally high at the boundary of the core. The density of the core starts at 12.3 gm/cm^3 and is much higher than at gm/cm^3 at the centre. Again, the density is increased to 13.6 in the depth of 6671 km. Along with that the gravity also increases with the increasing density and pressure in the earth's interior. Due to increasing pressure, the earth's substances remain in a stable and flexible state, not in a fully heated fluid; otherwise, the earth's interior will be floating with the liquid materials.

Table 2: Co-relationship between depth, density and atmospheric pressure

Depth (in km)	Density	Pressure (Atmospheric)
0	2.8	-
33	3.3	9
700	4.3-5.5	260
2,895	10.0-12.3	1,350
5,150	13.3	3,340
6,371	13.6	3,700

3
Isostasy

Various physical features, such as mountains, plateaus, plains, lakes, ocean basins are standing on the surface of the earth, holding a special balance between them. Without such balance, they cannot manage to remain in their present state. Whenever the balance between different parts of the earth is disturbed, different phenomenon such as earthquakes, volcanic eruptions or different geological events starts to occur.

The famous scientist Archimedes first adopted the concept of Isostasy. This theory shows that the volume of water needed to float on a freight vessel is much higher than that of an empty ship floating on the sea. This situation can be illustrated with the help of Fig. 6.

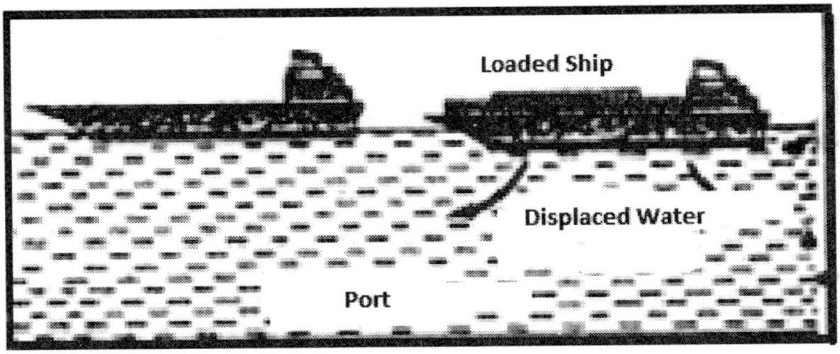

Source: Mitra, Kalpana, Sen, Jyotirmoy & Sengupta Purabi: *Physical Geography* (2014) Kalyani Publisher, Ludhiana, p. 72.

Fig. 6: Floating Condition of a Cargo ship.

If a region on the earth is located above the seabed, it is assumed that the area can be found at the bottom of the earth. In fact, the highlands and lowlands of the earth are standing in equilibrium state with themselves. For example, in the highest part of the Himalayas, the height of Everest (8848m) can be mentioned. However, it is true that the root of the Everest does not exist in the interior of the earth. It can be seen that although there is not that much space in the interior of the earth, it has root at the Sial level, and its main root remains present in the molten surface at the Sima. This situation is shown in the Fig. 7.

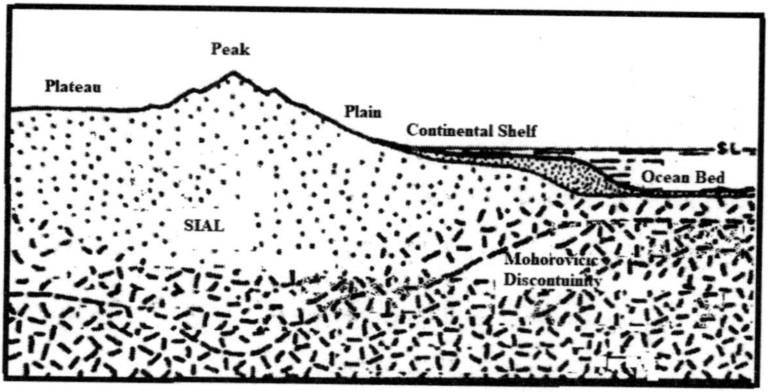

Source: Singh, Savindra: *Physical Geography* (2010), Prayag Pustak, Allahabad, p. 78.

Fig. 7: Airy's concept of Isostasy.

The Greek word "Isos" means "Equal" and "Stasis" means "standing still". **Isostasy** means the equilibrium position of earth between two units. The American geologist C.E. Dutton (1889) first used the word "Isostasy". He thought that the upper part of earth, such as mountains, plateaus, plain etc. remained in equilibrium state with the lower reaches of the earth's interior. According to him light continental layer (Sial) is floating over heavy oceanic layer (Sima).

Introduction to the Isostasy Theory: Although, the principles of equilibrium were first expressed in the contemplation of geologists, later the thought of attracting the gravitational pull of huge mountains. Sir Isaac Newton, inventor of the gravitational law hypothesized that the rate of attraction varies slightly with latitudes because the earth is not round completely. To examine this fact, in 1735, the French Academy sent two review teams to Lapland. A scientist named Pierre Bouquer, during his expedition to the Andes; he noticed that the upper part of the high Chimborazo peak is not attracting the plumbing line in any way. From that incident, he came to the fact that the gravitational acceleration of any area decreases with increasing height per meter. But, again, the rate of attraction that is located above the sea level is slightly increased at an area of high land surface. This effect on gravitational acceleration is termed as "Bouguer Mass effect."

Survey in favour of Isostasy theory

From the above idea, during the subsequent geodetic survey, Sir George Everest (1859) specified three places on the middle line of the Ganga-Brahmaputra plain region of India, namely (i) Kalyana at the foot of the Siwalik hills, (ii) Kalyanpur, 375 miles south of Kalyana, and (iii) Damriddaga region of the Deccan Plateau, 425 miles south of Kalianpur. There he measured the angle with the transit Theodolite and traced the horizontal plane with the vertical hanging plumbing line. Based on the vertical values of that plumb line, he determined the latitude and distances between them, and surveyed it in a triangular manner and calculated the distance and latitude of Kalyana and Kalyanpur. From that equation, the difference of the longitude of two places is about 5.23″. The reason for that was attributed to the location of the Himalayan Mountains. During the survey ran from the foothills of the Himalayan Mountains, it was observed that the Himalayan Mountains were attracting the plum line of Theodolite towards the Northward direction.

Concept of Sir George B. Airy (1802-1892): Although there are various debates about the principle of Isostasy theory, it is possible to squeeze an idea from those debates. The density of rocks of the upper part of Himalayas is much lighter than its lower part. Then gravity was measured in different parts of the earth and the same kind of inconsistency is observed. From these circumstances, scientists concluded that there is a difference between the rocks density of continents and ocean. **Sir George Airy** was the Astronomer who put forwarded his views on Isostasy in 1855. He stated that the continents are made of lighter **sial** which is floating on denser **sima**. In other words, it is said that the sial rock is floating on the sima. **J. Asteany** (1961) stated, "It is a boat in water sank into the denser material. In other words, Himalayas were not just a surface feature but lighter rocks of the mountain extend as roots up to considerable depth into the denser sima. Following the **Law of Floatation**, he argued that the high rising

Himalayas could be stable only if its larger part extended deep down into the denser rocks below.

Airy illustrated his view by taking the example of an iceberg. When the iceberg floats on the sea, its one part is above the water level and nine parts are immersed in water. The ratio of these floating conditions is 1:9. Depending on his theory, if the height of the Himalayas is 8,848 meters, the depth of Himalayas will be about 70,700 meters. At the same time, Airy also thought that the higher the position on the surface, the deeper its depth, i.e. "If the land column above the substratum is larger, its greater part would be submerged in the substratum and if the land column is lower, its lower or its smaller part would be submerged in the substratum". At the same time, Airy thought that the density of the various lower zones would always be the same, i.e. all the continents of the earth on which all the rocks are formed have the same density, only the difference between their height or their parallel position.

Airy proved his theory with the help of wooden blocks. He took wooden blocks of different sizes placed in a vessel containing water. He found that each block moved vertically and independently. As all blocks were made of wood, they had equal density but their heights were different from each other. It is seen that the blocks emerged by amounts, which were proportional to their heights, so that the longest block goes to the greatest height above the water and it also entered, deepest into the water. These blocks are said to be in a state of **hydrostatic equilibrium**. Again, Airy repeated same experiment with copper blocks placed into mercury and similar results were obtained. From this experiment, Airy argued that like wooden or cooper blocks, land features of the earth surface composed of rocks having similar density but they entered deep into the substratum in proportion to their height, i.e. the landmasses such as mountain, plateaus, basins and sea floor are almost in the same situation.

These experiments of Airy received considerable respect and admiration among the scientific community but in the later stage, different scientists noticed some weaknesses in the theory. As mentioned earlier, the height of Mt. Everest is 8848 meters and it has roots penetrating up to 70,700 meters (70.7 km) in the substratum. It has also proved that the temperature rises inside the earth at the rate of 1°C per 32 meters. The depth of Mt. Everest, which will be so high that the root cannot remain in a solid state. However, by accepting Airy's doctrine as the primary information, all scientists have acknowledged that the light substances (Sial) are floating on heavy materials (Sima).

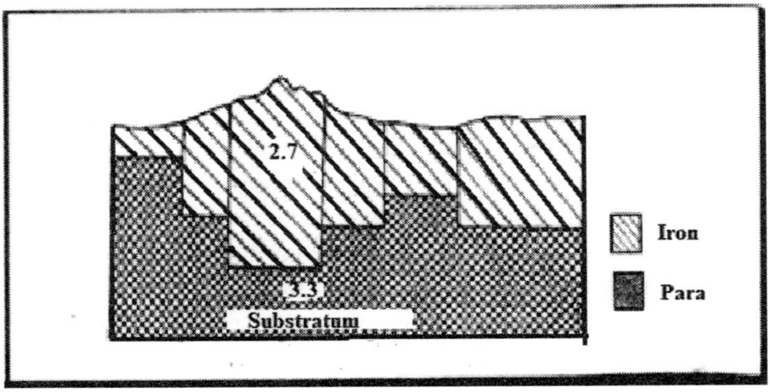

Fig. 8: Airy's concept of Isostasy.

Concept of Archdeacon Pratt: Pratt put forwarded his concept of Isostasy in the year 1855. After determining the cause of the longitudinal difference between Kalyana and Kalyanpur, the matter was investigated by Pratt. He worked on the assumption that the average density of the mountains was 2.75, approximately that of the continental rocks as a whole. He found that if these were the case, the difference should have been much greater, 15.885. In other words, the mountains were not exerting the attraction they ought.

Pratt further examined that the average density of the plateaus is higher than the average density of the mountain. Again, the average densities of plateaus is lower than the plains and the rocks of plains lower than the ocean floor. From this situation, he concluded that the increase in height decreases the density. In other words, there is an inverse relationship between height and density.

He also noted that there is a horizontal surface inside the earth where substances are uniform. That horizontal surface is known as **"Level of compensation"**. Above the level of compensation, the density does not change within one column but it changes from one column to the other. Thus, the central theme of Pratt's concept of isostasy may be exposed as uniform depth with varying density. According to him, equal horizontal surface must underlie equal mass along the level of compensation. Pratt supposed that the continents and other masses stand higher than the average because less dense material beneath compensates them. He imagined that some distance downwards was level at which the solid crust gave place to an elastic condition. Before Pratt exposed his opinion, by placing blocks of different materials (lead, irons antimony and zinc) in a mercury container are designed in such a way that they have equal weight. Since all the blocks have equal weight, they stand at uniform depth inside the mercury although their heights above the mercury level are different. Blocks of lower density will stand higher while blocks of higher density will be at lower level. From this experiment, Pratt argued that like these blocks, different physical features of the earth surface have a similar relation with the substratum.

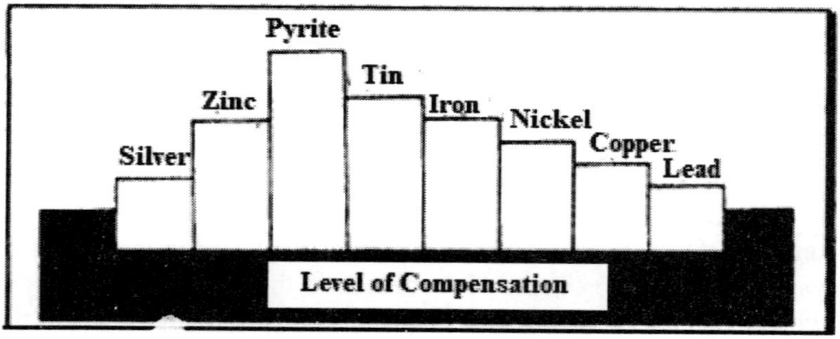

Fig. 9: Pratt's concept of Isostasy.

Several scientists examined various experiments and noticed that due to the varied density of rocks, the rate of gravity is varied from place to place. According to Pratt, density varies only in the lithosphere and not in the pyrosphere and barysphere. In order to prove the authenticity of this information J.M. Hayford & W. Bowie started geodetic survey along the coastal region of America.

Concept of Hayford and W. Bowie: After the Geodetic survey of the U.S. Coast, the both scientists gave a statement that the density of the earth surface is much lesser than the ocean floor and there is a flat, plain surface inside the earth called **"Level of compensation".** The views presented by Hayford and Bowie are more or less similar to those presented by Pratt. According to them, the **"Level of compensation"** exists in the 100 km. depth and they found that the variation of gravitational pull in that particular area decreased to 1/10th. Hayford also assumed that for contiguous columns of equal cross-section based upon level of compensation, the density was inversely proportional to the height. They were the first who assumed that crustal densities varied with elevation above this plane; under the mountains, the crust is lighter than under the oceans. Below the plane of complete compensation, density is uniform in all lateral directions.

Hayford proved his theory, taking five imaginary columns. The imaginary columns such as, plain, mountains, plateaus, coastal plain and off-shore region, which are reaching down to the level of compensation. He assumed that varying volume of matter in these columns is compensated by their density in such a way that they exert equal downward pressure at the level of compensation.

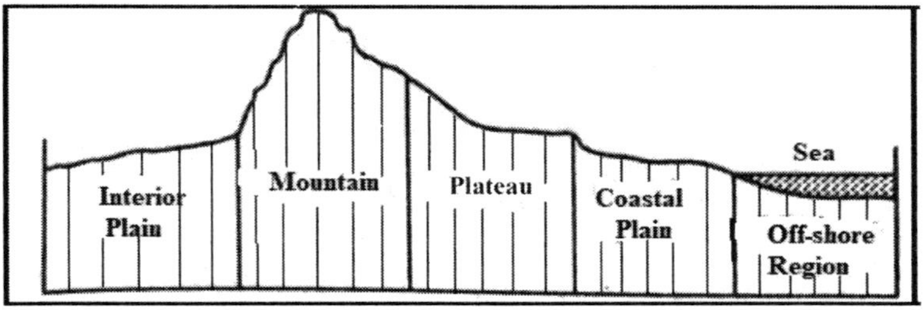

Fig. 10: Hayford's concept of Isostasy.

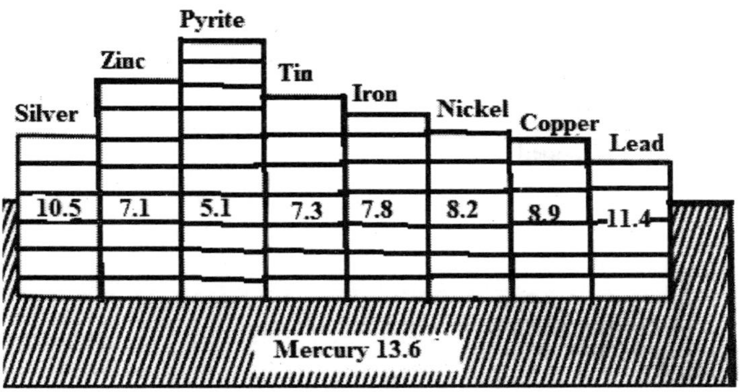

Source: Singh, Savindra: *Physical Geography* (2010), Prayag Pustak Bhawan, Allahabad, p. 77.

Fig. 11: Bowie's concept of Isostasy.

Bowie also exemplifies the concept of Apostasy by a transparent experiment. With the help of eight pieces of silver, zinc, pyrite, copper, lead, tin, iron, and nickel, Bowie tried to illustrate this concept. All those blocks had equal width and thickness but their heights varied from each other when all eight pieces were placed in a vessel containing mercury, it was found that in the bottom of all the eight pieces rest at the same level, but their height above the mercury is different from each other. This experiment proves that there is equal mass under equal surface area.

Isostatic Adjustment: It is true that the Isostatic adjustment is not possible at all because of the earth surface is never stable. In the course of time, new shape has been created on the surface of the earth, and the trend of this change on the earth has never stopped for a moment. This transformation of the surface is taking place in a varied manner. This movement takes place sometimes slowly or suddenly. Thus, even if there is disturbance of the equilibrium on the surface then there is a tendency towards equilibrium on the surface.

For example, when an uplifted landmass or mountains range is denuded by the erosion work of the agents of change, the load on the underlying column of the crust is reduced by the weight of the rock waste that have been carried away. At the same time, a neighbouring column, underlying a region of delta and sea floor where the rock waste is being deposited, receives corresponding load.

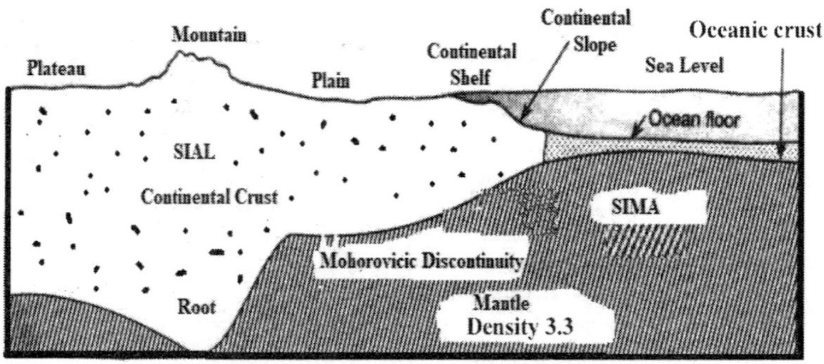

Source: Singh, Savindra: *Physical Geography* (2010), Prayag Pustak Bhawan, Allahabad, p. 78.

Fig. 12: Relationship between surface features and crustal structure (Arthur Holmns and Doris L. Holmes, *Physical Geology*, 1978, p. 21).

Briefly, when erosion begins at upland or at the different part of the mountain, the upper part of the terrain begins to unloaded, and the sediment accumulates in the near surface and creating pressure on the lower part. In response to this pressure difference in the mantle, a slow migration of material is set going. This process, hereby restored, is called **"Isostatic adjustment".**

From the above discussion, the following are the things that can be explained as follows—

1. The main driver of the combination of all types of Isocratic adjustment is the force of gravity.

   For example –

   (a) Erosion of hills and mountains results sediments.

   (b) Due to the reduction of the load of land surface, gradually elevated.

   (c) Deposition of sediments in the Delta region resulted into denudation land surface.

2. Very small loads, such as only a few hundred meters deep, are thought to be sufficient to accommodate stability.

4
Earth Movement and Landforms

Various theories about the creation of the earth has been shown to indicate that the earth was once a burning gaseous body and that the heated gaseous body gradually cooled due to radiation of heat and finally took the present shape of the earth. Due to radiation of heat produces, heavy pressure in the Earth's interior and causes many disastrous events in the earth's surface. These disasters are called **"diastrophism".** Again, these landforms creating forces are called **'tectonic forces'**. The word **'tectonic'** is derived from the Greek work **'Tecton'**. The Greeks used to call **Tecton**, the god of architecture, who created various sculptures. These whole processes are called **"Earth movement"**. In other words, the powerful internal forces operating from within the crust are called **"earth movement".** As a result, fold, fault, collision, fractures, slopes, earthquakes, volcanoes etc. are occurring on the surface of the earth. These processes of internal forces are mainly divided into two, as it is very sudden and gradual.

Long period changes that take place very gradually over a long period. People in general may not be easy to understand the trend of change.

On the other hand, **Short period changes or rapid changes** occur so abruptly that within a few seconds. For example, earthquakes and volcanic eruption, landslides are the forces, which change the landscape within a very short span of time. The forces affecting landscapes can be classified into two types –

1. Endogenetic force, and
2. Exogenetic force.

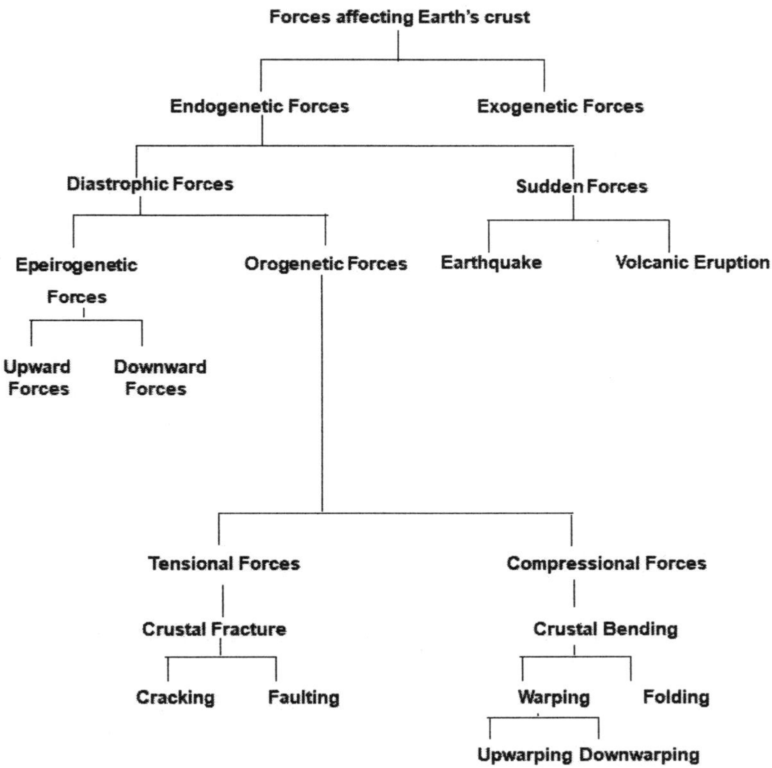

Fig. 13: Forces affecting crust of the Earth.

In this chapter, we will discuss about the effects of endogenetic forces.

Endogenetic Forces: The forces generated in the interior of the earth are called **endogenetic force**. Endogenetic forces works on the earth's surface in two directions—

1. Vertical Movement, and
2. Horizontal Movement.

Significant changes have been taking place on the surface of the earth under the influence of endogenetic forces. Mountains, plateaus, plains etc. are formed under the influence of vertical and horizontal movement. Although different scientists have given various explanations regarding endogentic forces, but it is true that the earth's interior is the source of this energy. Due to the change of heat and pressure inside the earth, cracks, faults etc. are created in different rock strata. Several times, this change occurs so fast that it creates a tremor within the interior. Endogenetic forces can be divided into two divisions –

1. Sudden force, and
2. Diastrophic force.

Sudden force: The Sudden energy is originated mainly from the center of the interior of the earth. This resulted in major changes in the interior and the changes are so sudden and rapid that no special action can be taken on the surface of the earth. As a result, severe damages to plants, animals and densely populated areas. Thus, Geologists have called it a catastrophic disaster. This is of course, creates a variety of landscapes on the surface. Therefore, sudden forces are also called constructive forces. For example, volcanic eruptions create various types of landscapes such as, volcanic cones, volcanic lakes or the plains of Deccan can be mentioned, as well as the various faults, lowlands, lakes etc. caused by the earth movement. Earthquakes create faults, fractures etc.

Diastrophic forces or Movements: Mountain building activities took place on the earth at different geologic past and are episodic in character. It is surmised that between orogenies there persist a long duration of relative stability in the earth's crust and subject to slow epeirogenic uplift or subsidence of the crust. However, there persist some controversies regarding the nature of occurrence of both orogenic and epeirogenic movement. Certainly, such contrasting views regarding the nature of diastrophism will profoundly differ in certain geomorphic interpretation and analysis.

Influence of earth movement usually results in (i) Horizontal, and (ii) Vertical changes on the earth's surface. These forces act very gradually and their effects become visible after thousands of years. Diastrophic forces results in many changes in the form of terrain such as folding, faulting, uplift and depression, which is also known as constructive forces. This is how creation of mountains, plains, plateaus, several lakes etc. are occurring across vast areas on the surface of the earth. Diastrophic forces can be classified into two types—

1. Epeirogenetic Movement, and
2. Orogenetic Movement.

1. Epeirogenetic Movement or forces: The word "*Epeirogenetic*" is derived from the two Greek words **"Epeiros"** meaning **"continent"** and **"genesis"** meaning **"origin"**. These movements work vertically on the surface of the earth. It affects larger part of the continents and causes upliftment and subsidence of continental masses through upward and downward movements. Epeirogenetic movement can be further sub-divided into two type, i.e. (i) **upward movement,** and (ii) **downward movement**.

(i) Upward movement: upward movement causes upliftment of the entire continent or part thereof rises above the sea level and forms a plateau or block mountains. Again, such rise on the coast is called **emergence**.

(ii) Downward movement: If any part of the earth surface collapse due to the downward movement, it is called **subsidence**. Likewise when the land area near the coast is subsides, it is called **submergence**. The islands of British Columbia were once part of the mainland. But due to submergence in the middle part, they have been separated from the mainland.

The result of the epeirogenetic movement is usually the creation of ancient plateaus, ridge valley, Block Mountain etc. Epeirogenetic movement is also responsible for the creation of ridge valley of Jordan and East Africa, the valley of Rhine and the Narmada-Tapi valley of India.

2. Orogenetic forces or movement: The word "**orogenetic**" is derived from two Greek words, "**oros**" meaning "**mountain**" and "**genesis**" meaning "**origin**". Thus, orogenetic forces or movements are associated with mountain building processes. Lateral or horizontal movement due to orogenetic forces create differences in pressure on the rocks in different parts of the earth. Horizontal forces are also called **tangential forces**. They work in two ways (i) Pressure from both directions. This is called **compression forces**, and (ii) In opposite directions due to pulling force. This force is known as **tensional force**.

Compressional forces cause **crustal bending** which are responsible for the formation of **folds**. On the other hand, tensional forces are associated with the creation of **crustal fracture, rupture and cracks and faults**.

When horizontal forces works face to face in the formation of fold, the rocks undergo the process of **"crustal bending"** in two ways (i) warping, and (ii) folding. In this case, the rocks are raised upwards or downward by force. The part that rises above the force is called **"up warping"**. On the other hand, the downward part is called "**down warping".** When the process of up warping and down warping of crustal rocks occurs in larger areas, it is called **"broad warping"**. In this way, compressive forces or movement causes bunckle and bend of crustal rocks, the resultant mechanism is called **"folding".**

Folds: Because of compression force on the surface of the earth, the pressure created on the crustal rocks causes the gradual upliftment or descendent of an area are called **fold**. Here, some parts are bent up and some are bent down. The upper part of the fold is called the **"anticline"** and the lower part of it is called **syncline**. The two sides of the fold are called **"Limb"**. Again, the middle part of the two limbs is called the **middle limb**. If a line is drawn between anticline and syncline, then it is called **"axis"**. The plane that divides the axis into two equal parts is called the **"axis plane"**. Based on anticline and syncline these axial planes are called as **axis of anticline** and **axis of syncline** respectively. Dip and strike give us the proper understanding of the structural form of rock beds. **Dip** is the inclination of rock beds with respect to horizontal plane. It includes both the direction of the maximum slope down bedding plane and the angle between the maximum slope and the horizontal. The dip direction is measured its true bearing. In this case, the compass reading is corrected for magnetic variation. Its angle is measured by the instrument known as **"clinometers".** The strike is the direction of any horizontal line along a rock-bedding plane. Strike always stands at the right angle to the Dip. For example, if the strike is east-west, Dip could be either to north or south direction.

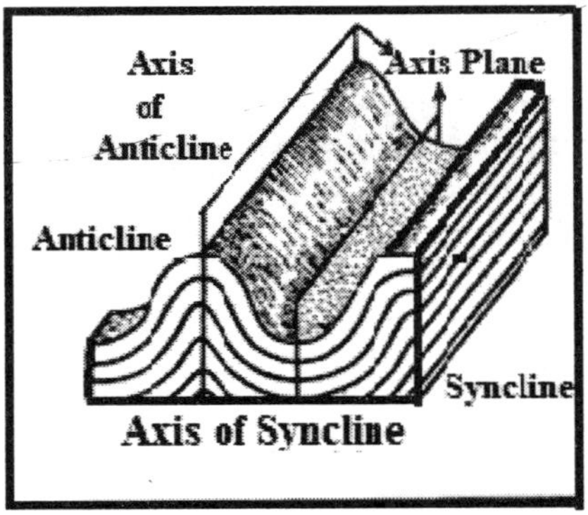

Fig. 14: Folding.

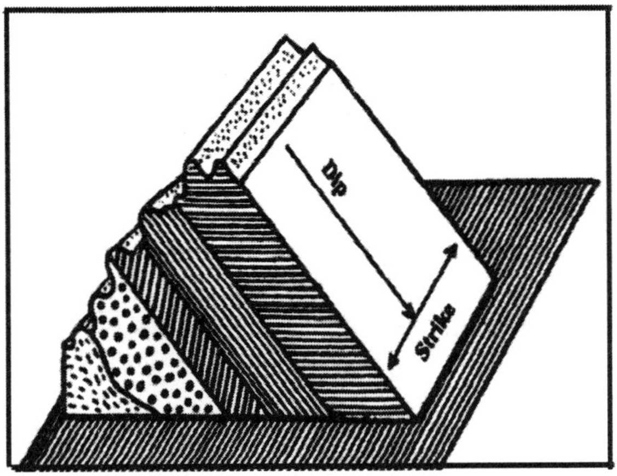

Fig. 15: Dip and Strike.

Source: Singh, Savindra: *Physical Geography* (2010), Prayag Pustak Bhawan, Allahabad, pp. 83-84.

Anticlinorium: It is a series of several tiny synclines and anticlines, which form in an extensive area. These are the result of horizontal compressive tangential force, which acts irregularly. Due to the varied intensity of compressive force such structures are formed.

Synclinorium: Synclinorium represents an extensive syncline having numerous minor synclines and anticlines. Such structure is formed due to irregular folding caused by irregular compression force.

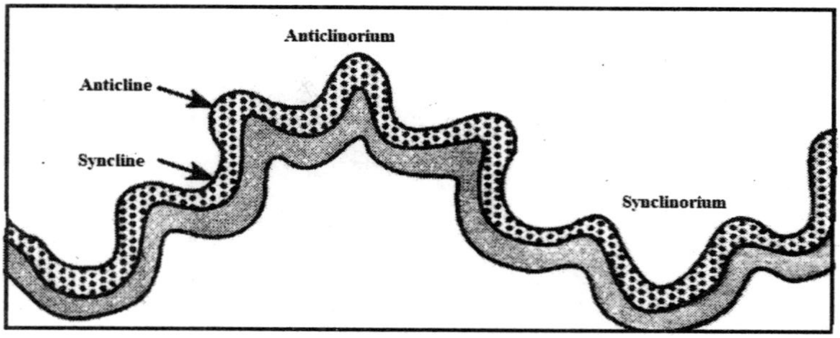

Source: Singh, Savindra: *Physical Geography* (2010), Prayag Pustak Bhawan, Allahabad, p. 84.

Fig 16: Anticlinorium and Synclinorium.

Types of fold: Depending upon several factors such as nature of rocks, and intensity of compressional force, different types of folds are formed. In this case, the effect of pressure on rocks is not uniform everywhere. Therefore, there is a difference between the nature and shape of the fold. According to the nature and structure of a fold, they are divided into several types—

1. Symmetrical folds: When both limbs on either side of a fold are equally bent at almost equal angles it is **called symmetrical fold**. These folds are an example of open fold that are rarely seen. Due to the equal compressive force on both sides of the limb of a fold, give birth to asymmetrical folds. The **Zagros Mountain** of south-west Iran is an example of a symmetrical fold.

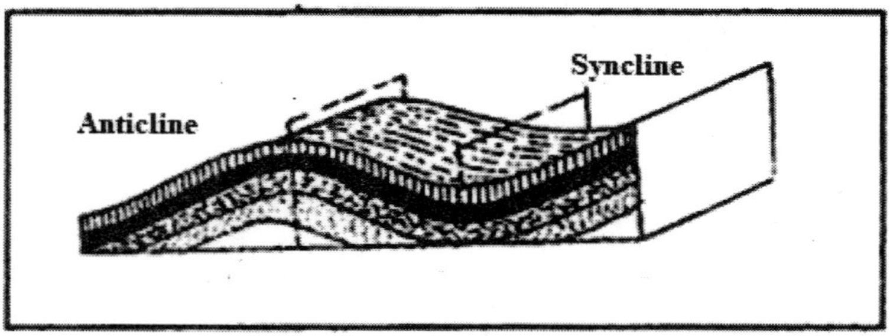

Source: Mitra, Kalpana, Sen, Jyotirmoy & Sengupta Purabi: *Physical Geography* (2014) Kalyani Publisher, Ludhiana, p. 106.

Fig. 17: Symmetrical fold.

2. Asymmetrical fold: These are formed when both limb of a fold are inclined at an unequal angle. Here, one limb is comparatively larger than the opposite limbs. In other words, one side of the fold is long and gradual in slope while the other is short and steep. This type of folding occurs if the impact of compressive force is greater on either one side; the intensity of the force gradually decreases downward.

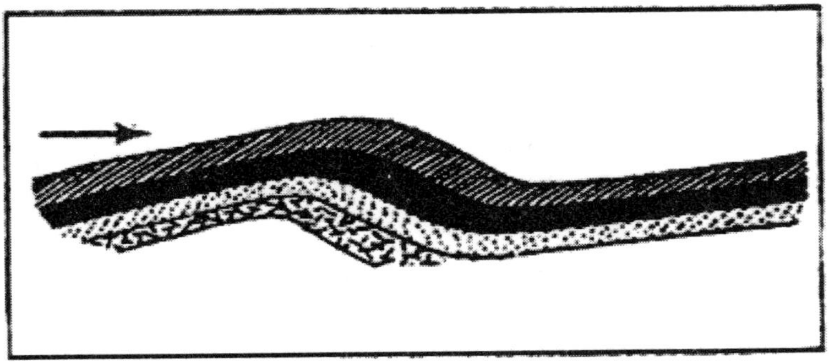

Source: Mitra, Kalpana, Sen, Jyotirmoy & Sengupta Purabi: *Physical Geography* (2014) Kalyani Publisher, Ludhiana, p. 106.

Fig. 18: Asymmetrical fold.

3. Monoclinal folds: When the compressional force acts in such a way that one limb inclines moderately while the other is vertical at right angle is called **monoclinal fold**. Unequal horizontal compressive forces are responsible for the creation of monoclinal fold.

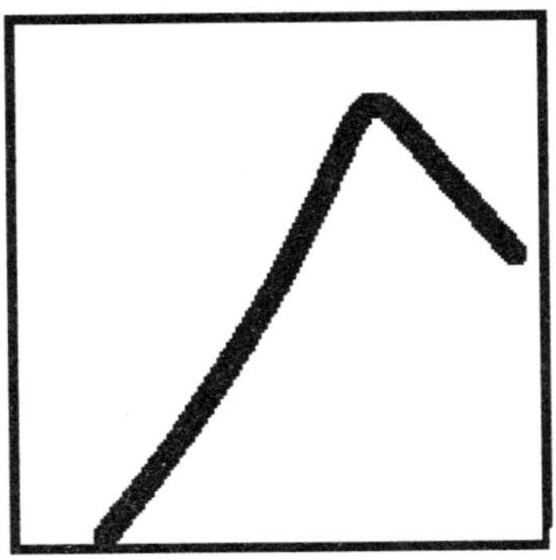

Fig. 19: Monoclinal fold.

4. Isoclinals folds: When the both limbs of a fold are parallel to each other at the same angles, they are called **parallel folds** or **isoclinals folds**. In these cases, the compressional forces are so strong that both the limbs of the fold become parallel rather than horizontal.

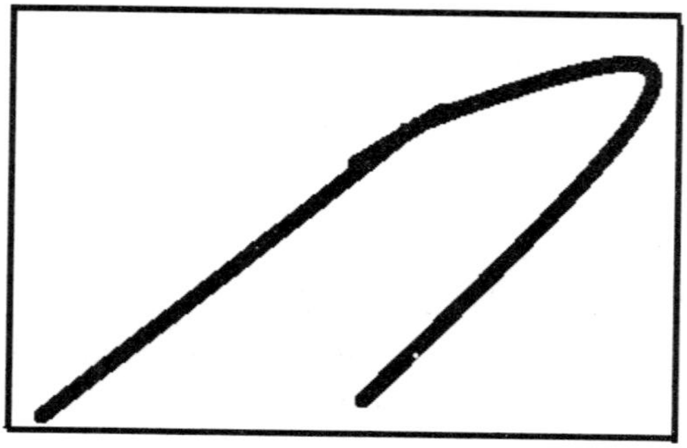

Fig. 20: Isoclinal fold.

5. Recumbent folds: When the axis plane and limbs of a fold are both in roughly horizontal positions and the compressive force of horizontal pressure increases abruptly, then one limb of the fold is pushed over the other limb. Such complex folds are called **recumbent fold**.

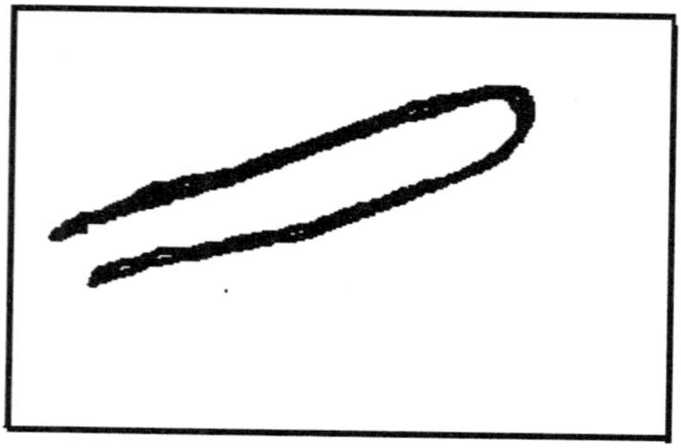

Fig. 21: Recumbent fold.

6. Overthrust folds: If the intensity of horizontal pressure or compressional forces increases on a recumbent fold, the rock get break in the middle of that fold and one part moves away from the other. This type of folding is called **overthrust fold**. The cracks along which the folds are separated are called **thrust plane**. In this case, limbs are seldom horizontal.

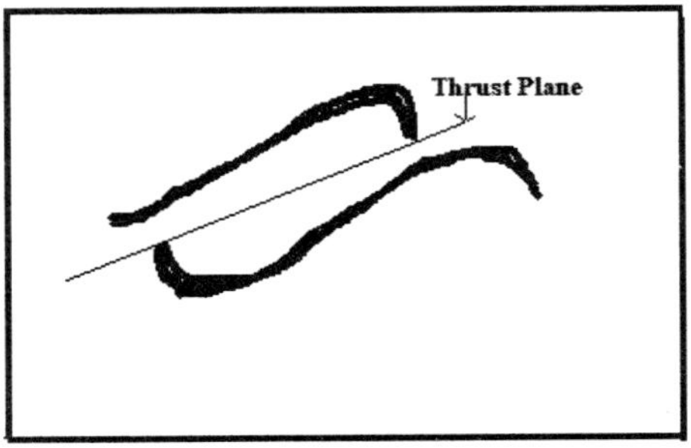

Fig. 22: Overthrust fold.

7. Plungs folds: When the intensity of compressive forces and their vibration in the earth's interior increases unevenly, the axis of the fold inclined at a fixed angle rather than in parallel with the surface, this type of fold is called **plungs fold**.

8. Fan folds: Such type of folding is formed by the compressional force causes extensive and broad fold consisting of several small folds from both sides. Consequently, the structure of which resembles the shape of a fan. Hence, it is known as **fan fold**. Such type of fold can be seen in the Alps mountain system.

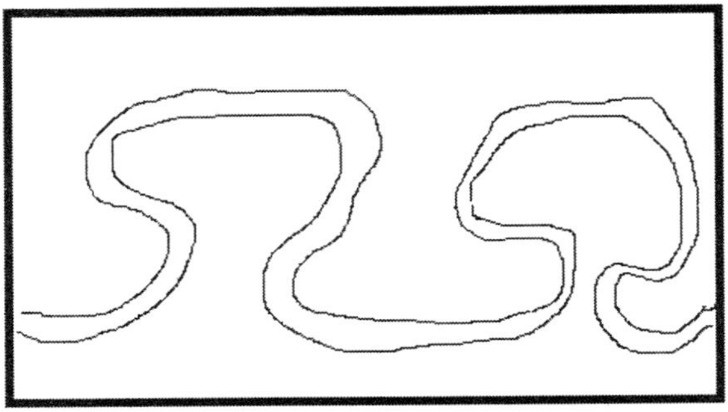

Fig. 23: Fan folds.

9. Open folds: Due to the moderate nature of the compressive force, such folds are formed. The angle between the two limbs of the fold is 90°-180°.

10. Close folds: When compressive forces are intense, the limbs (both) of the fold are positioned almost vertical. As a result, the "anticlines" are seen to be very close to each other or angle between the two limbs of a fold is acute angle. This type of folds are called **close fold**.

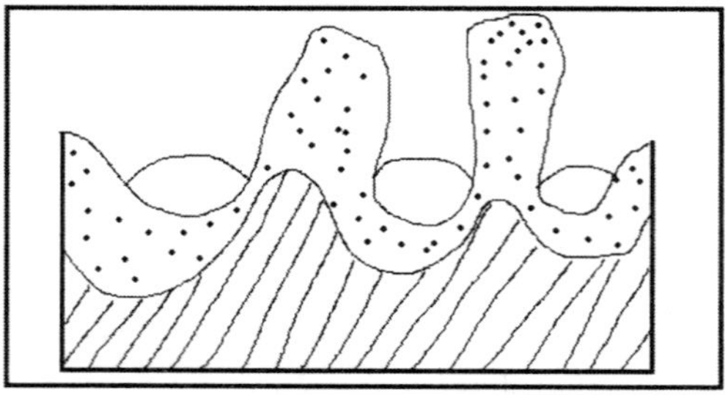

Fig. 24: Close folds.

Nappe: The word **Nappe** is derived from French word, which means **sheet** such as tablecloth. It is a large size recumbent fold. Due to enormous horizontal pressure, one limb of the recumbent folds slides forward and overrides the other fold and thrown several km. away from its original place, they are called **nappe**. Nappes are mostly found in the Alps mountain range. Several Nappes also can be seen in the western Himalayas. In Himalayas, Nappe has been discovered by Wadia from the Kashmir Himalayas. (Fig. 25)

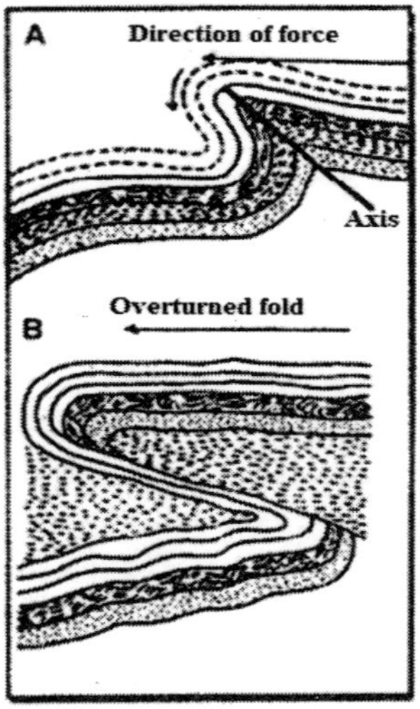

Source: Mitra, Kalpana, Sen, Jyotirmoy & Sengupta Purabi: *Physical Geography* (2014) Kalyani Publisher, Ludhiana, p. 107.

Fig. 25: Nappes.

Crustal Fracture: Both tensional and compressional forces are responsible for crustal fracture. In other words, displacement of rocks along a plane due to tensional and compressional force is known as **crustal fracture**. They may either act horizontally or vertically or even both ways. Crustal fracture is always depending on the intensity of tensional forces and nature of rocks in that particular area. Weak or moderate tensional forces cause only cracks in the rocks; but strong tensional forces may cause dislocation of rocks that may lead to the formation of a **fault**. Crustal fracture is generally divided into two types (i) Joints, and (ii) faults. A Joint is only a crack in the rocks. The cracks become a fault when there is complete dislocation of rocks on both sides of a fracture and parallel to it.

Faults: As the effect of compression and contraction on the rock, strata formed various types of fold, in the same way strong vibrations or movement in the earth's interior causes deep crack in the earth's crust and one part of the crack separates and diverges from other, which is termed as fault. In other words, rapture is a part of land broken and its one part moves forward or backward or upward or down-ward more than the other part. According to *Oxford Dictionary of Geography*, "fault is a fractured surface in the earth's crust along which rock have travelled relative to each other." The *Oxford Dictionary* gave an example in this regard. That is, the Tecton faults in the US state of Wyoming is about 50 km long. The faults have a vertical distance of about 6000 meters. Usually many faults occur simultaneously. They sometimes face each other or cross each other at different angles.

About the origin of the fault, it can be said that the endogenetic forces in the earth's crust causes the compression and contraction on the rock strata and expansion takes place in several parts of the crust. In this case, compression and contraction process is convergent and expansions are divergent process. Under the effect of convergent pressure, the crust ruptures into many parts. This formation process of rupture is called **Block faulting**. When tensional force exceeds, the tolerance capacity of the rock strata, the crust splits into many pieces. The plane along which the fault occurs is called the **"fault plane"** and the line at which the fault intersects the surface is called the **"fault line"**.

Fault Terminology: A number of terms associated with the ideal fault which should be understood properly before going into the details of the formation of different types of faults—

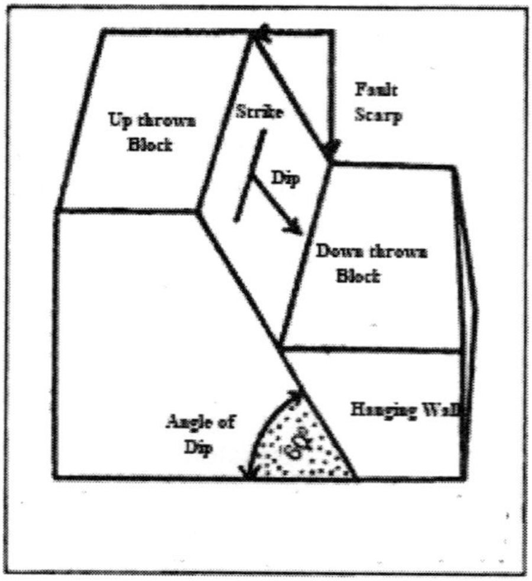

Source: Singh, Savindra: *Physical Geography* (2010), Prayag Pustak, Allahabad p. 87.

Fig. 26: Fault.

Fault Plane: Fault plane refers to the plane or surface along which the rock strata dislocated due to tensional and compressional forces acting horizontally, vertically or at different angle to rock strata. A fault plane may be vertical, horizontal, smooth, inclined, or curved.

- **(i) Fault Dip:** Fault Dip is the angle between fault plane and horizontal plane.
- **(ii) Fault strike:** Fault strike refers to the fault plane in precisely the same way as to the rock-bedding plane.
- **(iii) Upthrown side**: Upthrown side represents the uppermost block of a fault.
- **(iv) Downthrown side**: It represents the lowermost block of a fault.
- **(v) Hanging wall:** The upper wall of the fault is called Hanging wall.
- **(vi) Footwall:** It represents the lower wall of the fault.
- **(vii) Fault scrap:** It is the steep wall like slope formed by faulting of crustal rocks.

Type of fault: Based on origin and nature, different types of faults are recognized. Some of the important faults are described as under—

(i) Normal fault: These types of faults are formed due to the tensional force that cause the blocks on either side of the fault plane move in opposite direction. Generally, the angle of normal fault is between 65°-90° or vertical. A normal fault leads to the formation of steep and cliff like feature is called fault scarp. Generally, fault scarp varies from few km to more than hundred km, and its height and the length covers up to 300 km.

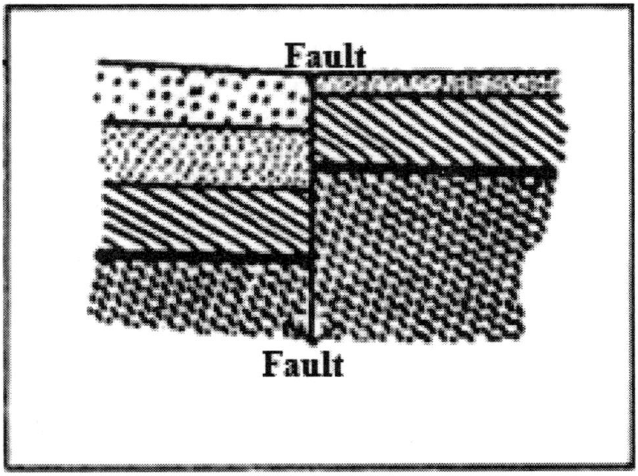

Source: Mitra, Kalpana, Sen, Jyotirmoy & Sengupta Purabi: *Physical Geography* (2014) Kalyani Publisher, Ludhiana, p. 117.

Fig. 27: Normal Fault.

(ii) Reverse Fault: Excessive compressive pressure causes one part of the fault plane rises above the other part is called **reverse fault**. In the case of reverse faults, the blocks along the fault plane move towards each other. The fault segment seems to be standing upright like a hanging cliff. The hanging cliff thus formed often fails to carry their load and cause landslides. If the hanging rock cliff is bent too much on the downward side, is called **thrust fault**. This type of fault plane has a dip angle of less than 45°. The main boundary fault between Shiwalik and the middle part of lower Himalaya is a prime example of this type of fault. Also, since reverse fault raises above the other due to excessive pressure and horizontal movement, it is also called **"compression fault".** Most of the time when one part of the fault rises above the other due to excessive pressure, then it is called **"Overthrust**

fault". In this case, the angle of this fault plane is less than 10°. This type of fault can be noticed in the Himalayas and Alps.

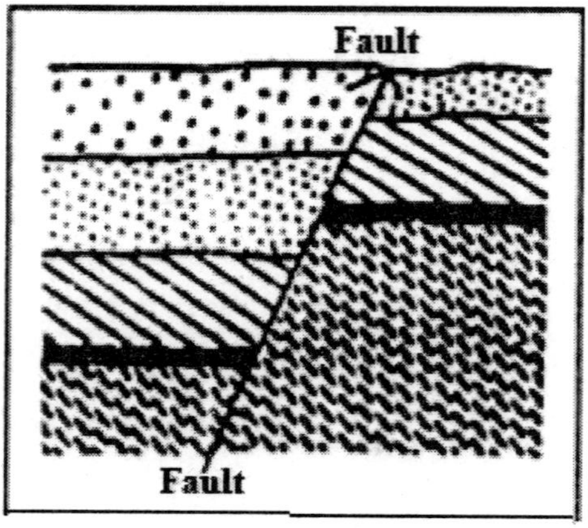

Source: Mitra, Kalpana, Sen, Jyotirmoy & Sengupta Purabi: *Physical Geography* (2014) Kalyani Publisher, Ludhiana, p. 118.

Fig. 28: Reverse Fault.

(iii) Lateral or Strike slip fault: When rock strata displaced in a horizontal direction along the fault plane and follows the strike line is called **strike slip fault or lateral fault**. The angle of inclination of lateral fault is 0°. The angle is based on the angular position of the horizontal displacement of the two rock block along the fault plane. If the displacement occurs to the left from the side of the fault is called

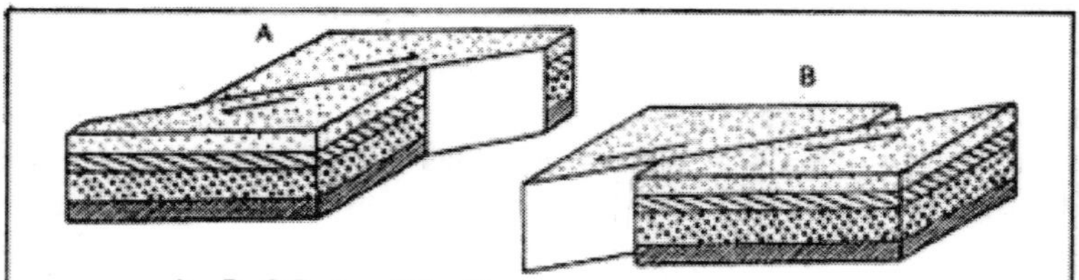

Source: Singh, Savindra: *Physical Geography* (2010), Prayag Pustak Bhawan, Allahabad, p. 89.

Fig. 29: Lateral fault.

left lateral fault and when its sift occurs from right is called **right lateral or dextral fault**. If the fault plane is located almost parallel to the rock bed layer is called **bedding fault**.

(iv) Step faults: When a series of faults are created parellely in the same direction, the resultant faults are called **step faults**. In this case, several normal faults are created together and the faults gradually go down towards the bottom like stairs. This type of fault can be seen in the rift valley of plane sides in Europe.

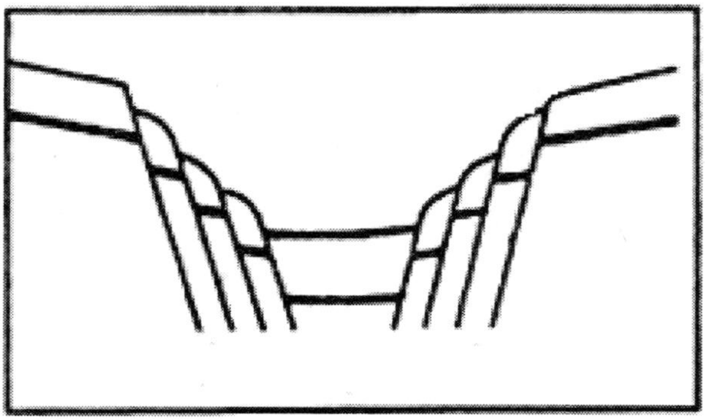

Source: Mitra, Kalpana, Sen, Jyotirmoy & Sengupta Purabi: *Physical Geography* (2014) Kalyani Publisher, Ludhiana, p. 120.

Fig. 30: Step fault.

(v) Rift Valleys or Grabens: Faulting activities are responsible for the creation of rift valley. Rift valleys are formed due to displacement and subsidence of crustal part between the two normal faults. They represent a trough and depressed areas between two crustal parts due to vertical or horizontal movement. The shapes of the rift valleys are long and narrow troughs. Trough referred to fault blocks, which are depressed below the adjacent upthrown blocks. The rift valley is usually found between the two Block Mountains. In German, the rift valleys are generally called **"grab-en".** Most of the geologists have proved that the river Narmada flows through such a rift valley between the Vindhya and Satpura Mountains of India. The rift valley is formed in two ways—

(i) The parts between the two fault sites and the land surface on both sides of the fault is in high portion.

(ii) The parts between two faults are fixed for a long time but the land on its two sides rises upwards, which creates rift valleys.

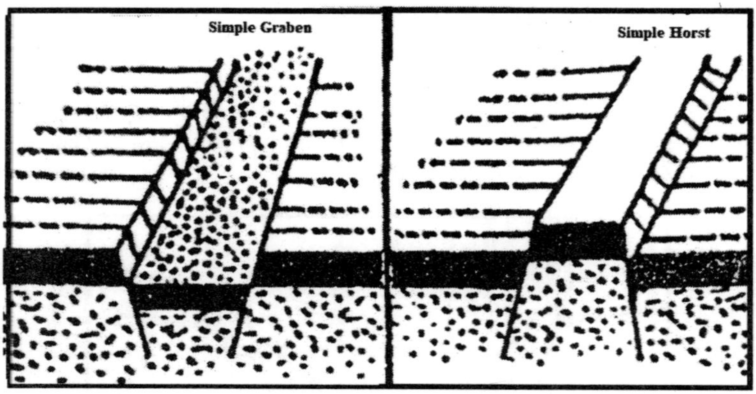

Source: Singh, Savindra: *Physical Geography* (2010), Prayag Pustak Bhawan, Allahabad, p. 89.

Fig. 31: Rift valley or Graben.

The rift valley is usually long, deep and short in width. The best example of such a long valley is the **Jordan Valley** or **the Red sea Basin** to the **Zambese basin**. This valley is about 4,800 km width. On the other hand, some of the valleys are so deep that their depth touches the depths of the sea. The **Death Valley** in Southern California is a prime example of such valley. Rift valleys are not only seen on the surface but also formed at the bottom of the sea. Such deepest Grabens exist as trenches or ocean deeps.

5
Plate Tectonic Theory

The rigid lithospheric slabs or solid crustal layers of the earth are called **plates**. These plates are experiencing movement in horizontal direction, which results in breaking and bending of the crustal rocks. This whole mechanism is called "**Plate Tectonic**". Several theories that are prevalent and established among the scientific community in the present age regarding the formation of the earth's landform, i.e. mountains, plateaus, continents, spreading of sea floor, volcanoes etc. Among them, the theory that was established in 1967 is the **"plate tectonic"** theory. In connection with this theory, on the one hand, there is Wagner's **Continental Drift Theory**, on the other hand, the theory of **Sea Floor Spreading**, perhaps to explain the geometric form of the relative expansion of the oceanic crust. Plate tectonic theory is able to describe the formation, location etc. of earthquakes, mid-Atlantic ridge, transform fault creation of new mountains, islands of Pacific Ocean, Deep Ocean trenches etc. Like other doctrines, it is not based on only one thought provoking information. In fact, it is based on practical experiments by varied scientists. As a result, it is no longer judged as information but as a scientific fact. Among the scientists who experimented with this subject were Tuzo Wilson, B. Isaks, J. Oliver, L.R. Sykes, Mekenzie, W.J. Morgan, Paker, X.Le. Pichon and J.R. Heistzler.

Basic concept of the plate tectonic theory: Scientist believe that earth's lithosphere is broken into about a dozen of plates, which slide by, collide with, or separate from each other as they move over earth's interior. These plates are created where they separate and are recycled where they collide in a continuous process of creation and destruction. Continents, embedded in the lithosphere, drift along with the moving plates. The theory of plate tectonic describes the movement of plates and the forces acting between them. It also explains the distribution of many large-scale geographical features—mountain chains, structures on the sea floor, volcanoes and earthquakes that result from movements at plate boundaries.

Location on plate on the earth crust

According to X. Le. Pichon (1968), there are 7 major plates, 9 minor plates and 19 very little plates standing on the asthenosphere zone. Of these, seven large plates of different sizes are particularly significant.

(a) Major Plates

(i) Pacific plate,
(ii) Eurasian plate,
(iii) North American plate,
(iv) South American plate,
(v) Australian plate,
(vi) African plate, and
(vii) Antarctica plate.

(b) Minor Plates

(i) Arabian plate,
(ii) Bismark plate,
(iii) Caribbean plate,
(iv) Carolina plate,
(v) Cocos plate,
(vi) Juan-de-Fuca plate,
(vii) Nazca or East pacific plate,
(viii) Phillippines plate, and
(ix) Scotia plate.

(c) Very little plates

(i) Persian plate,
(ii) Anatolia plate,
(iii) China plate, and
(iv) Fiji plate.

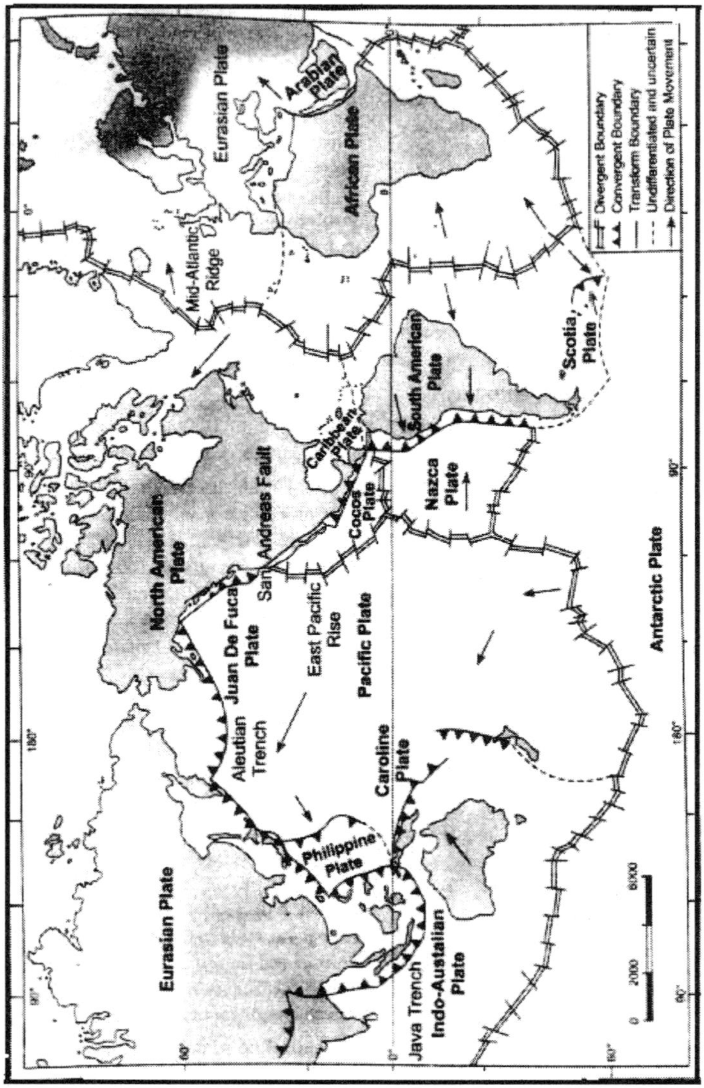

Fig. 32: World map of lithospheric plates. (Khullar, D.R., *Physical Geography*, 2014, p. 157)

The central parts of these plates are seemingly stable and free from the effects of earth movement, but their edges are always moving, and scientists used the location of different seismic centers to determine the boundaries of these plates. Various experiments have shown that the earthquake prone regions and major volcances are located across long stretches of land along the boundaries of the plates. Although there are some volcanoes in the middle of some of the plates but their number is negligible.

Plate boundary or Margin: The edge of the plates are called **plate boundary**. In fact, it has been observed that the centers of all earth movement activities are located along the border of the plates. Observing the plate boundaries shows that there are three main motions at these boundaries—

(i) Diverging or constructive plate boundary,
(ii) Destructive or converging plate boundary or collision plate boundary, and
(iii) Conservative plate boundary.

(i) Diverging or constructive plate boundary: In this case, one plate moves away from the other. At divergent boundaries, where plates move apart, partially molten mantle material rises and fills the gap between them. This material forms new lithosphere added to the irregular edges of the diverging plates.

On the seafloor, the boundary between separating plates is marked by a mid-ocean ridge that exhibits active basaltic volcanism, shallow-focus earthquakes, and normal faulting caused by tensional or stretching forces created by the pulling apart of two plates. The process by which plates are separated and ocean crust is created is called **seafloor spreading**. The Mid-Atlantic Ridge and the East pacific Rise are spreading centres that have created millions of square kilometers of seafloor.

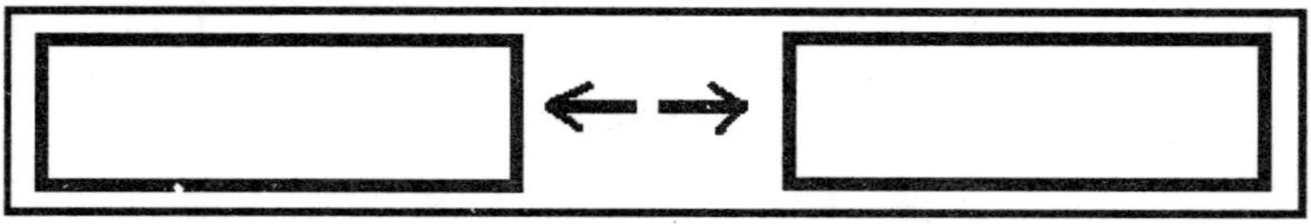

Fig. 33: Diverging plate boundary.

On the other hand, in early stages of plate separation can be found on continents. Such sites are characterized by long, down faulted rift valleys, basaltic volcanic activity and shallow-focus earthquakes.

The Great Rift Valley of East-Africa between the African plate and the Somali sub-plate is thought to represent an early stage of plate separation within a continent. Sometimes, continental rifting may slow down or stop before the continent splits apart and a new ocean basin opens up. The East African Rift valley and the Rhine valley in Europe, both of which are still mildly active, may be in such places. The Red Sea and the Gulf of California are rifts that are further along in spreading, because of their greater width and depth which have been flooded by the ocean. The Arabian Peninsula is splitting away from Africa at the Red Sea and Baja California is separating from the Mexican mainland at the Gulf of California.

(ii) Destructive or converging plate boundary: Plates collide along convergent boundaries. The overridden plate is subducted, or thrust downward into the mantle below, where ultimately it recycled. Collision and subduction produces deep sea trenches, adjacent mountain ranges of folded and faulted rocks, and magmatic belts. The magmatic belt can be a mountain range on land or a chain of volcanic islands, called an **island arc** on the seafloor. Once subducted lithosphere is heated, water and other volatiles "boil off". These boil hot materials in the wedge of mantle above and induce melting. This is the source of magma that feeds and builds the overlying chain of volcanoes. The collision of two plates generates very large forces in the region, and in a general way, these forces must result in the faulting that triggers the shallow and deep focus earthquakes that occurs in subduction zones. By the time, the lithosheric plate has move from a divergent to a convergent boundary, it has cooled to become denser than the mantle below.

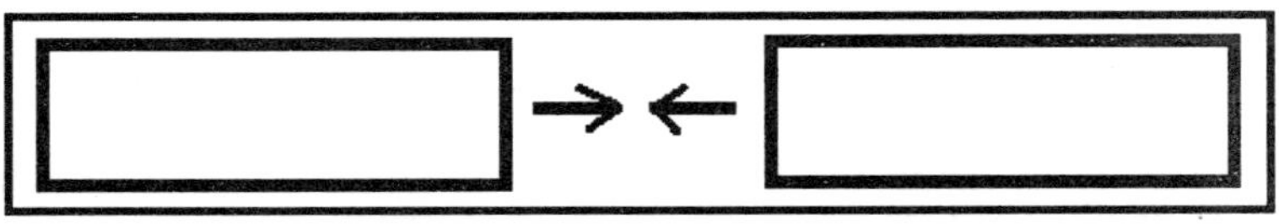

Fig. 34: Converging plate boundary.

A convergent boundary that exhibits all of these features is the one between the Nazca plate and the South American plate. The Peru and Chile, which are situated at the offshore of deep sea trench, the Andes Mountains with their many volcanoes and some of the world's largest shallow and deep focus earthquakes occur in such places.

When a plate collision involves between two continents, both tend to remain afloat. The collision of India and Asia plate is a good example. In this case, the Eurasian plate is overriding the Indian plate, creating a double thickness of crust and forming the highest mountain range in the world, the Himalayas. The Himalayas are supported isostatically by a crustal root projecting into the denser mantle below.

(iii) Transform Boundaries: At transform boundaries, plates slide past each other neither creating nor destroying lithosphere. Transform faults occur where the continuity of a divergent boundary is broken and offset. **The San Andreas fault in California**, where the pacific plate slides by the North American plate, is a prime example of a transform boundary on land. Shallow focus earthquakes with horizontal slips occur on transform boundaries.

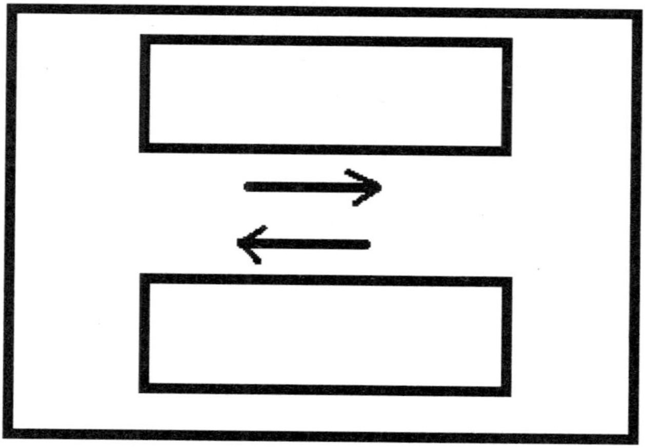

Fig. 35: Converging plate boundary.

Combination of plate Boundaries: Each plate is bounded by some combination of these three kinds of boundaries. For example, the Nazca plate in the pacific is bounded on three sides by zones of divergence, along which new lithosphere forms, and on one side by the Peru-Chile subduction zone, where lithosphere is consumed. Most of the North American Plates are bounded by the divergent Mid-Atlantic Ridge on the east; the San Andreas Fault and other transform boundaries on the west and zones of subduction and transform plates run from Oregon to the Aleutians on the north-west.

Causes of plate Movement: It will be difficult to understand plate tectonic theory completely until we can explain why plates move. The International Geodynamics project, a global program of coordinated experiments, enlisted the efforts of thousands of scientists in many countries to discover the underlying causes of plate motions.

1. Convection current: The information that geologists have gathered through seismometers and the conclusions reached by them are that the current originate in the interior of the earth's crust rises to the surface of the earth. Because of this convection current, the liquid of the earth's crust rises to the surface and sinks to the bottom in a cyclic way through the cool and hard rock. This circulation directly or indirectly controls the movement of plate due to convection current.

2. Pushing mechanism: When the pressure of high molten material below the plate boundary increases and its presence creates tremendous pressure on the plate boundary. This pressure increases to the outward on both sides of the plate boundary and causes the plates to move in opposite directions.

3. Pulling Mechanism: When two plates are close to each other due to the effect of an earthquake, the small plate sinks to the bottom of the other. As a result, the surrounding plates are also attracted towards the boundary. This causes movement between the plates.

4. Oceanic crust formation and Rates of motion: In the case of Ocean floor spreading, it has been observed that there is a fine coherence of ocean floor spreading extending from both sides of the ocean floor due to the effect of convection current. Scientists have observed that just as the expansion of the oceans is 1-5 cm per year and in the same way, the destruction of the ocean floor at the rate of 5-15 cm per year, which is not possible without the movement of the plates.

5. Force of Hot spot: Scientists have speculated that the release of magma from all the hotspots on the Earth's crust may have resulted in plate motion.

Criticism: The plate tectonic theory is also facing a lot of criticism—

1. This theory does not give an exact explanation of how the motion of the plate was originally generated.
2. No evidence has been found on the existence of convection currents in the asthenosphere, but it is thought that the deepest

core of the asthenosphere is the source of the convection currents.

3. The force needed to move the plate was also not found.
4. In extension, the length of the continental ridge is much longer than the length of the area when in the plate tectonic theory.
5. On the other hand, looking at the extent of the oceans, it can be seen that all the oceans are expanding and plates are being destroyed, but this theory cannot explain why Pacific Ocean surrounded by almost all parts of the volcanic zone. The penetration of large granite rocks to the west of North America has pushed the continent west. On the other hand, the north-westerly movement of the continent along the San Andreas Fault line has not been able to explain this theory.
6. Although it is possible to explain the formation of some mountain with the help of plate tectonic theory, but there is no correlation between the formation of hills in the eastern part of Australia or the formation of the Drakensberg Mountains in south Africa where upliftment of land are primarily responsible.
7. At the end, although the theory of the position of the plates is self evident, this theory does not explain how the plates are formed.

In conclusion, it is not possible to arrive at accurate information about the various complex processes in the interior, as it is not possible to enter into the interior of the earth. At the same time, it is clear that the movement of the plate is not a process. All the processes together help in the movement of the plate and the heat of the interior conducts all these processes. At present, there is no need to separate the theory of continental drift because these two theories have proven in the light of the plate tectonic theory. With this theory, it is now possible to explain the structure and shape of continents and oceans, as well as the causes and effects of earthquakes and volcanoes.

6
Earthquakes

"An earthquake is a motion of the ground surface, ranging from a faint tremor to a wild motion capable of shaking buildings apart and causing gaping fissures to open in the ground. The earthquake is form of energy of motion transmitted through the surface layer of the earth circles from a point of sudden energy release—the focus." (A.N. Strahler and A.H. Strahler, 1976). Intensity of earthquake is usually measured by the Richter scale which was devised in 1935 by Charles, F. Richter. The scale is devised to measure the intensity of earthquake within the range of 0 to 9. As the scale is a logarithmic, therefore the scale has no upper limit of number. The 1934, Bihar earthquake measuring 8.6 Richter scale are among the greatest earthquake so far recorded in the world.

Earthquakes are originated at the interior of the earth and the point of origin is called 'focus'. The point on the surface of the earth vertically above the focus is called 'epicentre'. The wave generated by an earthquake are known as 'seismic waves', these waves are recorded by a devise called 'seismograph' or 'seismometer'. 'Seismology' is the termed used to study the seismic waves. The earthquake releases three types of waves (i) P-wave or longitudinal waves, (ii) S-wave or Transverse waves, and (c) L-wave or surface waves.

The movement, velocity and travel time taken by these waves are measured by 'seismograph'. From the observations it is ensured that the higher the density of the rocks through which they travel or pass, the greater will be the speed of the waves. The study on the earthquake waves reveals that up to the depth of 1800 miles or 2850 km, the speed of the waves increases progressively. Beyond that depth of 1800 miles, the speed of the waves changes in a peculiar manner. Usually, P-wave and S-wave move towards the interior of the earth while the L-wave travels only over the surface of the earth. At the depth of 1800 miles of the earth's interior the state of matters changes from solid to liquid and for that condition the velocity of the waves changes. Primarily S-wave is found absent beyond the depth of 1800 miles as the whole belt is composed of liquid materials. Only the P-wave although penetrates into this liquid layer of the earth's interior but its velocity is reduced and S-wave is deflected back to the surface of the earth. The L-wave, which travels over the surface of the earth causes all the damages to most of the structure of the earth.

Measuring of earthquakes: People have always tried to quantify the size of, and damage done by earthquakes. Since early in the 20th century, there have been three measuring methods of earthquakes. The oldest of the scales is called the Mercalli intensity scale. Earthquakes are described in terms of what nearby residents felt and the damage that was done to nearby structures. This scale is qualitative in information because it is based on visual damage and not the actual energy released by the earthquake.

Table 3: Measuring of earthquakes based on Richter scale

Magnitude	Description	Intensity	Frequency
Less than 2.0	Micro	Normally only recorded by Seismographs. Most people cannot feel them.	Million per year
2.0-2.9	Minor	A few people feel them. No building damages.	Over one million per year
3.0-3.9	Minor	Some people feel them. Objects inside can be seen shaking.	Over 1,00,000 per year
4.0-4.9	Light	Most people feel it. Indoor objects shake or fall to the floor.	10,000-15,000 per year
5.0-5.9	Moderate	Can damage or destroy buildings not designed to withstand earthquakes. Everyone feel it.	1000-1500 per year
6.0-6.9	Strong	Widespread shaking far from the epicenter. Damages buildings.	100-150 per year

(*Contd.*)

Magnitude	Description	Intensity	Frequency
7.0-7.9	Major	Widespread damages in most areas.	10-20 per year
8.0-8.9	Great	Widespread damages in large areas.	About once per year
9.0-9.9	Great	Severe damages to most buildings.	Once per 5-50 year
10.0 or over	Massive	Never recorded.	Never recorded

With the invention of seismograph station, the **Richter magnitude scale** was created which was developed by **Charles Richter** in the year in 1935. This scale uses a seismometer to measure the magnitude of the largest jolt of energy released by an earthquake. Today, the **moment magnitude scale** has replaced the Richter scale. The moment magnitude is calculated from the area of the fault that is ruptured and the distance the ground moved along the fault. The Richter scale and the moment magnitude scale are logarithmic.

Distribution of Earthquake

Earthquakes are unevenly distributed on the surface of the earth. In some places they occur rapidly and very destructive, in other regions their effect is hardly perceptible and the rate of their occurrence is very low. The most outstanding aspect of the distribution of earthquake is their occurrence in narrow continuous belts (Fig. 36). This pattern of earthquake distribution is associated with the weaker and isostatically distributed areas of the globe. Most of the world's earthquakes occur in following zones:

(i) Zones of young folded mountains,
(ii) Zones of faulting and fracturing,
(iii) Zones of representing the junction of continental and oceanic margins,
(iv) Zones of volcanic activity, and
(v) Along different plate boundaries.

There are three well-defined belts or zones of seismic activity in the world, and these are:

(a) The Circum-Pacific Belt,
(b) The Mediterranean and Trans-Asiatic zones, and
(c) The Mid-Atlantic ridge and the African rift system.

(a) The Circum-Pacific Belt: This belt encompasses the coastal margins of the Pacific Ocean. It is the most active seismic belt of the world and accounts for about 65% of the total earthquakes of the world. This belt presents four ideal conditions for the occurrences of earthquakes viz. (i) junctions of Continental and Oceanic margins, (ii) zones of young folded mountains, (iii) zones of Active Volcanoes, (iv) Subduction zone of destructive plate boundaries. The earthquake associated on the eastern coastal margins of the Pacific Ocean is starting from Alaska and runs towards the south, parallel to the kurile, Japan, Mariana and Philippine trenches; beyond with it divides into two branches, one going west parallel to the Indonesia trench and the other towards the Keramadec-Tonga trenches to the north-west of New Zealand. In this belt the earthquakes are caused due to the collision of the Pacific and the Asiatic plates and consequent vulcanicity. Japan records about 1500 seismic shocks every year. On the western coastal margin of Pacific Ocean the earthquake zone of North America and South America represented by the Rockies and the Andes folded mountain ranges respectively. These zones are isostatically very sensitive zones because these are also the zones of convergent plate boundaries where the Pacific Oceanic plate is being continuously subducted below the American plates. In addition, these are the areas of intense volcanic activity. The recent earthquake of Mexico city in 1985 reveals the impact of collision of convergent plate boundaries on the occurrences of earthquakes. The damages done by devastating earthquake included death of about 5,000 people, disappearances of about 2000 persons, injuries to about 40,000 people, destruction of about 4000 buildings, damages to about 6000 buildings, lesser damages to around 50,000 buildings etc.

(b) The Mediterranean and Trans-Asiatic Belt: This belt is also known as the Mid-Continental belt on Alpine-Himalayan belt represents the collision or subduction zones of continental plates. This is the second most important seismic belt where about 21 per cent of the world's total seismic events are recoded. This belt extends along the Mediterranean Sea, the Alpine mountains and their offshoots in Europe, Northern Africa, Eastern Africa, Asia Minor and the Caucasus,

Iran, and Pakistan to the Himalayan mountain system including the Pamirs, Tibet and China to as far as the Myanmar hills and Indonesian islands. This belt represents the weaker zone of young fold mountains where isostatic and fault induced earthquakes are caused due to the subduction of African and Indian plates below Eurasian plate. This belt is characterized by longer and shallow foci earthquakes and also some intermediate foci earthquakes. Deep foci earthquakes are negligibly small.

(c) The Mid-Atlantic Ridges and African Rift system belt: This belt lies along the mid-Atlantic ridge and transform faults. The mid-Atlantic ridge is a major fracture zone where American plates and European plates diverge and constant upwelling of magma is forming new Oceanic crust. The spreading of sea floor and fissure type volcanic eruption cause earthquakes of moderate to mild intensity with shallow focus (less than 7 km deep). The rift valley of East Africa and Red Sea are considered as an extension of this belt.

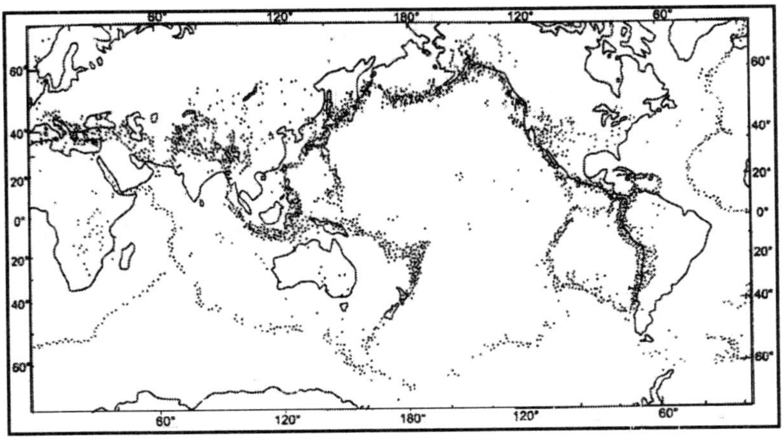

Fig. 36: Distribution of earthquakes in the world.

Causes of Earthquake

Disequilibrium at any part of the earth's crust causes earthquake in the earth. Disequilibrium or instant imbalance is caused by various factors like volcanic eruption, faulting and folding, up warping and down warping, gaseous expansion and contraction inside the earth, by drastic pressure caused by man-made water reservoirs and plate movement. The distribution of earthquake occurrence ensured that the earthquake belts are closely attached with the weaker zones and isostatically imbalanced areas of the world. Contrarily it is believed that isostatically stable and old stable rigid masses are free from any seismic disasters. The causes of earthquakes are discussed below in brief—

1. Vulcanicity

Volcanic eruption is associated with the occurrence of earthquakes. Volcanic activity earthquakes are so intimately related to each other as they acts like cause and effect for each other. Volcanic eruption gives rise to earthquake by the process of creating tremendous pressure on the surface of the earth. The explosive violent gases try to escape upwards and push the surface from below with tremendous force and results in the occurrence of earth tremor of huge magnitude. As the gases comes out, breaking the crustal surface they generate tremendous force and resulted in the disequilibrium in the crustal surface to cause severe earthquakes in an around their point of origin. Obviously, the magnitude of such earthquakes primarily depends on the intensity of volcanic eruption. Krakatoa volcano of Japan generates tremendous force and caused violet earth tremor the impact of which was felt in areas of Cape Horn located at a distance of 12,800 km.

2. Faulting and Folding

The formation of faults and folds are product of endogentic forces which acts upon the surface in both horizontal and vertical dimension. The resultant effects develops isostatic imbalance in the crustal rocks. Disequilibrium thus generated causes earthquakes of varying magnitude depending upon the nature and intensity of displacement of rocks blocks. Sudden displacement of rocks blocks induced the formation of folds and faults of varying size and more often caused earth tremors. For example, the 1950, Assam earthquake was believed to be the resultant disequilibrium in crustal rocks induced by crustal fracture. As such, the seismic events near Bhasta Dam and Koyna Dam of Maharastra are the resultant effects of active faulting beneath the Deccan tracts. N.K. Brahman and J.G. Niyogi, the scientists of National Geographical Research Institute opined that gradual increase in the seismic events in Bhasta Dam area since 1983 was the product of active faulting beneath the basaltic crust of the Deccan trap. Mainly the intensive faulting also caused Bhuj earthquake of 2001.

3. Hydrostatic pressure and Atmospheric causes

It is believed that the earthquakes are caused mainly by endogenetic forces but certain human activities also accidently cause earth tremors of serious consequences. Specific human activities such as underground mining, nuclear explosion, storage of large volume of water in reservoirs, blasting of rocks by dynamites for constructional works etc. also develops earth tremors. The storage of huge volume of water in the reservoirs causes isostatic imbalance, which in turn, results in the sudden earth tremors in the area. Numbers of major earthquakes are related with dams and reservoirs all over the world. For instance, the Greece earthquake of 1931 was the direct product of Marathon Dam constructed in 1929. Other such examples are earthquakes of U.S.A, in 1936 around Hoover Dam, Koyna earthquakes of 1967 in Maharastra, Monteynard and Grandvile earthquake of France, Mangla earthquake of Pakistan, Manic earthquake of Canada, Kariba earthquake of Zambia, Hendrick Verqoerd of South Africa, Kurobe earthquake of Japan etc. are all products of the construction of Dams and reservoirs which ultimately dismantle the isostatic equilibrium causing earthquakes in the nearby areas.

4. Plate Tectonic Theory

Plate tectonic theory recently has achieved widespread response for its explanation on the causes of earthquakes. According to this theory, the crust of the earth is composed of solid and movig plates in both continental and oceanic areas. A total of six major plates in the crust are identified. These are African plate, Indian Plate, Eurasian plate, American plate, Pacific plate, and Antarctic plate along with a total of twenty minor plates. These plates are not static and are continuously moving due to thermal convective currents originated at the earth's interior. As a result, the plate boundaries could remain stable for long period and the most conspicuous consequences are reflecting in all the tectonic events that mostly take place along the boundaries of moving plates. Following the nature of tectonic events, the plate boundaries are divided primarily into three types, these are (i) constructive plate boundaries, (ii) Destructive plate boundaries, and (iii) Conservative plate boundaries.

Constructive boundaries are those which represent the trailing ends of divergent plates moving in opposite direction from each other.

The destructive plate boundaries represents the two converged plates towards each other and collide to create tremors of high magnitude. Usually, the heavier plate subducts below the lighter plate boundary and thus crustal instability in that part of the earth's crust. On the other hand, the conservative plate boundaries are those where two nearby plates slip past from each other without causing any form of collision. Constructive plate boundaries give birth to faults and ruptures while the destructive plate boundaries give rise to intensive faulting and folding at or near the collision zone of the crust. Conservative plate boundaries sometimes results in the formation of transform fault along the plate boundaries only. Earthquakes of varying magnitudes are closely attached with these different type of plate movement.

Earthquake of high magnitude are often occurred along the destructive plate boundaries. The collision of two conversing plates and the subduction of heavier plate below the lighter one give rise to the events like mountain building, massive faulting and violent volcanic eruptions, which, in turn cause devastating earthquake. Maximum earthquakes are associated with the process of plate convergence. Most of the earthquakes occurred on the earth's surface are closely attached with the plate boundaries.

The intensity of earthquakes and their hazardous consequences are merely determined on the basis of quantum of damages done to human lives and prosperity. The magnitude of seismic intensity although is measured by Richter scale but the concern of damages caused to life and property are more often taken into consideration to assess the consequences of earthquake. The hazardous affects are assessed on the basis of damages caused to the human life, human property and the human constructions. It includes deformation of ground surface, destruction of human construction such as buildings, roads, railways, dams, bridges, destruction of habitats of both villages and cities, loss of human and animal lives and property, landslides, floods etc. The consequences of earthquake are summarized as: Earthquakes of high intensity caused severe damages to human structures such as buildings, roads, railways, dams, and reservoirs bridges etc. cause damages to human and animal life even.

The hilly and mountainous areas, which are, composed of weaker lithologies and tectonically instable faces severe landslides and debris falls as a result of earthquakes tremors. Slope instability and slope failure ultimately augment the process of landslide and debris falls. Landslides and debris falls ultimately damage human settlement and transport system on the downhill areas. It may be pointed out that in the surface composed of unconsolidated surface materials alluviums, colluviums, swamps deposits, levelled depression etc. reclaimed by dumping coarse sands and gerbages. The vibrations of earthquakes last longer and amplitudes of seismic waves become more intense than areas composed of unconsolidated materials and hardrocks. Therefore, the areas composed of unconsolidated materials faces severe damages in comparison to other areas.

Earthquakes causes severe damage to human structures like roads, railways, buildings, dams, bridges, etc. It is quite common that earthquakes cause severe damage in the areas unconsolidated surface than consolidated ground and stable bedrocks. With high density of buildings and large agglomeration of population, the cities are the worst sufferer from the effects of earthquake. As a result of earth tremors of higher magnitudes the large buildings are collapsed and men and women are buried under the debris of building materials, water supply pipes, electric poles, telephone wires etc. all are disrupted and such a situation leads to severe crisis of life. Transport systems also are collapsed and thus make the life more hazardous. The Tang-Shan earthquake of China (1976) records highest casualty of human life accounting more than 7,50,000.

Earthquake of high magnitude results in the total collapse of dams and reservoirs and cause severe flash floods in the downhill areas. Floods are also resulted from the blocking of water flow of rivers by the rock blocks and debris produced by the earthquake tremors on the hillslopes. Sudden removal of blockage causes severe flood in the down valley region and leads to the loss of life and property of the region. The 1950 Assam earthquake witnessed such devastating flash flood in the Dihang and Subansiri river in both upstream and downstream areas. Earthquakes originate in the sea or nearby areas generate high sea waves due to intense shaking of the ground and cause great loss of both life and property of the inhabitants of Seashore regions. The areas of Pacific Ocean are very much prone to frequent Tsunamis as the ring of earthquakes and volcanoes girdles the ocean.

Severe earth tremors caused by earthquakes create extensive deformation of ground surface. The frequent rise and subsidence of ground surface and faulting activity are the prime factors for the deformation of ground surface. Deformation ground surface induced severe damage to the life and property of people along with drastic changes in the ground conditions. The 1897 Assam earthquake caused a massive fault of 19.3 km long. As such the alluvial-filled areas of the flood plains of Mississippi river were fractured at many places and leads to the collapse of ground surface. Sometimes the strong vibrations generated by earthquakes intensively shake the buildings and other constructions and leads to the occurrence of severe fires. Such fires cause primarily from overturning of cooking gas cylinders, contact of live electric wires, churning of blast furnaces, displacement of other electric and wire attached appliances. Such a pathetic event took place in the vicinity of Tokyo in 1923 where more than 38,000 casualties out of a total of 1,63,000 was due to severe fire breaking. The Turky earthquake, 1999 caused a total devastation of two oil refineries.

7
Volcanoes

"A volcano is a vent or opening, usually circular or nearly circular in form, through which heated materials consisting of gases, water, liquid lava, and fragments of rocks are ejected from the highly heated interior to the surface of the earth." (P.G. Worcester, 1948). According to S.W. Wooldridge and R.S. Morgan (1959), "the term vulcanicity covers all those processes in which molten rock materials or magma rises into the crust or is poured out on its surface, there to solidify as a crystalline or semicrystalline rock." Volcanoes are those sudden and abrupt explosion of smoke, dust, lava, rock fragments which astound human understanding. It is indeed a line of communication between the interior mass of magma and the surface of the earth. The mouth of volcano is called 'Caldera'. Its mountain shape is due to the accumulation of lava comes out of it. The materials ejected by the volcano is usually liquid, solid and gaseous. The liquid materials ejected by a volcano is called lava and is composed of molten rocks and minerals.

Lava is of two types: (i) with high content of silica is called acidic lava, and (ii) with less content of silica is called basic or basaltic lava, the acidic lava is usually bright in colour, light in weight and melts relatively at a higher temperature. Certainly, the basic or basaltic lavas are darker in colour and heavier in weight. It melts comparatively at a lower temperature.

Component of Volcanoes

The Volcanic eruptions are caused by the formation of a massive tunnel in the rock layer of the earth's crust, which spreads all over the earth's surface and thus solidity around the cracks or tunnels, creating a mountain like an elevated ridge. The cracks through which underground lava flows out are usually round in shape and this path is called the **"volcanic pipe".** The mouth of the volcanic pipe that emits lava is called **"volcanic vent".** Generally, the shape of volcanic vent is U shaped. The mouth of a volcano is called **"crater"** and the vast size crater is called **"caldera"**. The diameter of this type of caldera ranges from one kilometer to 25 kilometers and where magma is stored is called the **"magma chamber".**

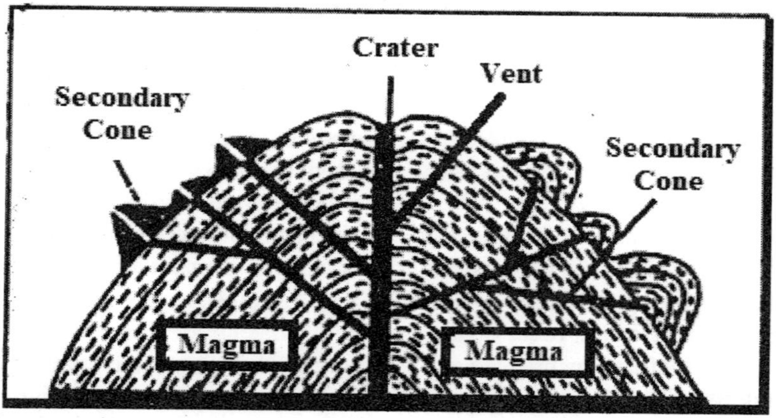

Source: Khullar, D.R.: *Physical Geography* (2014), Kalyani Publishers, Ludhiana, p. 206.

Fig. 35: Components of Volcano.

Classification on the basis of periodicity of Eruptions

Based on periodicity of Eruptions, volcanoes can be divided in three categories—

1. Active,
2. Dormant, and
3. Extinct.

1. Active volcanoes: Volcanoes that erupt frequently after eruption or have recently erupted are called **active volcanoes**. In such cases, the volume of the magma chamber of the volcano is usually large. However, in this case, the level of explosion is not always the same or all the flammable substances are emitted together. Therefore, active volcanoes are divided into two divisions according to the level of emission.

Active Volcanoes

Active Incessant **Active Intermittent**

(a) Active Incessant: Volcanoes that erupt nonstop are called **"active incessant"**. There are many active incessant volcanoes on earth. **Lasser peak of California** is the chief among them. Moreover, 13 such active incessant volcanoes are present in **Kamchatka**.

(b) Active Intermittent volcanoes: Volcanoes that erupt for several consecutive days are called **active intermittent volcanoes**. In some cases, after the eruption, when the pressure of the flammable material, including various gases, decreases in the magma chamber, the eruption stops for sometimes. When the pressure increases again eruptions takes place. **Mt. Etna** in Sicily, **Klyuchevskaya** in Kamchatka is one such volcano.

2. Dormant volcanoes: Volcanoes that have erupted in the past but are now inactive and carry signs of future eruptions are called **dormant volcanoes**. In such cases, once large amount of volcanic eruptions are emitted, the gas and pressure in the magma chambers decreases to such an extent that they become completely silent. More often, they become active again as a result of prolonged inactivity, that is, when the pressure in the magma chamber gradually increases. Such volcanoes are extremely dangerous. **Fujiyama** in Japan or **Krakatoa** in Indonesia is example of such type of volcanoes.

3. Extinct volcanoes: Volcanoes that have erupted in the past but there are no indications of future eruption are called **extinct volcanoes**. Extinct volcanoes are those whose activity has not been known to us during the historical time and it is very difficult to say whether a volcano is extinct or dormant. For example, the **Vesuvius**, the **Krakotoa**, **Mt. Unzen** of Japan and **Mt. Pintubo** of Philippines. There are about 158 big and small extinct volcanoes in the world. For example, **Kamchatka,** including **Datung** in China's Shanxi province and **Tentung** in Yunan province.

Classification on the basis of the nature of lava flow:

Based on the nature of lava flow, volcanoes can be classified into two types:

1. Central Eruption, and
2. Fissure Eruption.

1. Central Eruption: When lava comes out of a pipe or through a crack, it is called **central Eruption**. Depending on the nature and quality of gases, pressure, pryroclastic lava rocks etc. Central eruption type of volcanoes can be further sub-divided into following types—

(a) Hawaiian Type: Sometimes, the liquid lava flows far and wide along the slope of the volcano without any explosion. This is because of less viscous lava and violent nature of gases. This type of eruption occurs more frequently in the volcanoes of Hawaii Island located in the Pacific Ocean, so it is called **Hawaiian type**. The viscous lava, which are easily movable and has high temperature (1200°C). These volcanoes create crater and the lavas are always in molten state. If the gases released quietly, the lava will rise from a few meters to several hundred meters. However when the strong wind blows, these liquid lavas fly in the air like long glassy threads known as **Pele's** hair (Pele being the Hawaiian goddess of fire). As the amount of lava in a volcanic eruption increases, the lava pours down to the slope of the volcano. This type of eruption causes shield volcanoes.

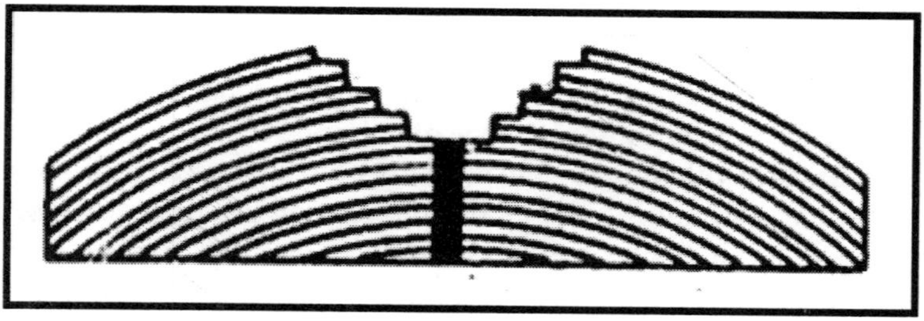

Fig. 38: Hawaiian type of Volcano.

(b) Strombolian type: Strombly, the volcanoes of Lipari Island, located in the north of Sicily. The lava here is mainly basalt and the temperature is between 1000°C -1100°C. These lava flows are relatively less mobile and contain a large amount of gases. Volcanoes erupt very rhythmically every few minutes to a few hours or every few days. As magma is less liquid, it rises above the gas, comes in contact with the cooling gases, solidifies and falls to the ground in the form of rock fragments and is embedded in **Volcanic Mountains**. The lava fountain activity of Strombli, reflected at night as a red glow on downside of a towering stream plume has earned the title of **"Lighthouse of the Mediterranean".**

Fig. 39: Strombolian type of Volcano.

(c) Vulcanian type: Vulcanian type of volcano located in the Lipari group of islands in Italy. Because of their lava is much viscous in nature, lava cannot spread farther down to the slope of the volcano. As a result, it quickly solidifies and seals the mouth of the crater in between two eruptions. During the second eruption, a lot of rock fragments and ashes are scattered in the sky and when these gases expand upwards, it creates a dark clouds much like **cauliflower**. These emitted ashes are transported everywhere under the influence of wind.

Fig. 40: Strombolian type of Volcano.

(d) Vesuvian type: Such eruptions are also comprised in the Stromblian and vulcanian type. Different types of lava such as intermediate andesite, andesitic basalt and andesitic desitic are erupted during the eruption. The lavas are viscous in nature and of low mobility. The temperature of the lavas are 800°C-1000°C or higher and these volcanoes frequently become active which is expressed in major eruptions, great gas explosions and ejection of ample of solid products. Another feature of this eruption is the formation of a large number of secondary volcanic cones on the sides of the volcano. There are about 200 such secondary cones in the Etna volcano.

Fig. 41: Vesuvian type of Volcano.

(e) Pelean Type: It is called so after the **Mont Pelee** volcano, which is located in the Island of Martinique of the Lesser Antilles. Andesitic lava is primarily released during the eruption of this volcano. These are very viscous and gas saturated. This viscous accumulates in the volcanic pipe, forming **dome**. The lava gas cannot escape from under that dome. As a result, its slope gradually increases. At one point, this gas was released by penetrating a weak spot in the corner of the dome and causing a huge explosion. These gases are mixed with burning ashes and various rocks particles. These are known as **"Nue ardent".** The lava flows down to the slope of the volcano at high speeds, causing severe damage to nearby settlement area. In 1902, almost all of the 30,000 people in the town of St. Pierre died because of such a Nue ardents flow from Mount Pele.

Fig. 42: Pelean type of Volcano.

(f) Krakatoa type: Krakatoa is an island located in the **Munda Strait** between Java and Sumatra. The island was created by three volcanoes. Gases and ashes-laden lava erupted violently from the volcano. In 1833, a catastrophic eruption occurred from this volcano. The eruption sound of this volcano also heard from Australia. The mixture of gases and ashes rose to a height of about 40 km, creating a cloud of blood-colored in the sky. The eruption caused two-thirds of the island submerged to the bottom of the ocean, which created a huge **Caldera**.

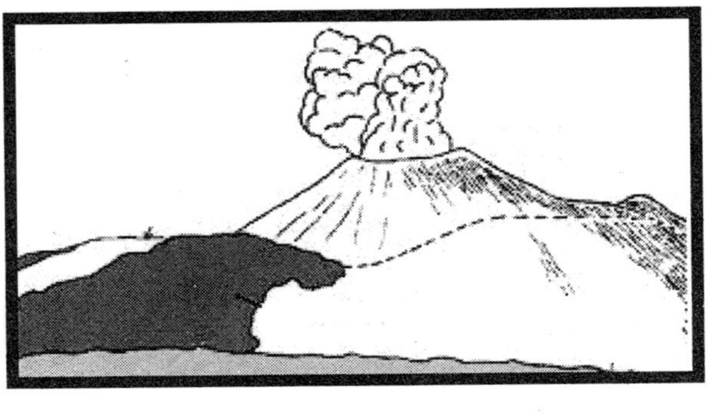

Fig. 43: Pelean type of Volcano.

2. Fissure Eruptions: In such an eruption, huge amount of lava comes out silently through various cracks of the earth's surface and flows along the slope of the volcano. In 1783, a lava flowed from Laki volcano which is 32 km long crater in Iceland, flooded an area of about 557 sq. km. This type of lava flow formed the **Colombian plateau** on the banks of river in North America and the **Deccan plateau** of South India. Similar eruptions can also be observed in islands such as **Hawaii** and **Samosa**.

World Distribution of Volcanoes: There are more than five hundred active volcanoes in addition to several thousand extinct volcanoes in the world. Like earthquakes, the spatial distribution of volcanoes over the globe is well defined and well understood because volcanoes are found is a well defined belt or zone (Fig. 44). If we see the world distribution of volcanoes, it shows that volcanoes generally occur in the weaker zones of the earth's crust and primarily associated with the seismic event like earthquakes. The weaker zones of earth are represented by folded mountains (Western Cordillera of North America, Andes mountain chain of South America, the Alps in Europe, the Himalayas in Asia) and fault zones. Volcanoes also occur at the meeting zones of continents and oceans. Occurrences of volcanoes along the coastal margins and during wet season indicate close relationship between water and volcanic eruption. Moreover, volcanic eruptions are also associated with the mountain building and fracturing. Based on plate tectonic theory, there is a close relationship between plate margins and vulcanicity and it has been found that most of the world's active volcanoes erupt at the plate boundaries. About 80 per cent of world's active volcanoes are found at convergent or destructive plate boundaries and about 15 per cent active volcanoes are associated with the divergent or constructive plate boundaries. Besides, some volcanoes are also found in intra-plate regions, e.g. volcanoes of Hawaii Island, fault zones of East Africa etc.

As mentioned earlier, volcanoes occur in specific zones or belts of which there are three major belts or zones of volcanoes in the world as like earthquakes zones, viz.:

1. Circum-Pacific belt,
2. Mid-Continental belt,
3. Mid-Atlantic belt, and,
4. Intra-plate Volcanoes.

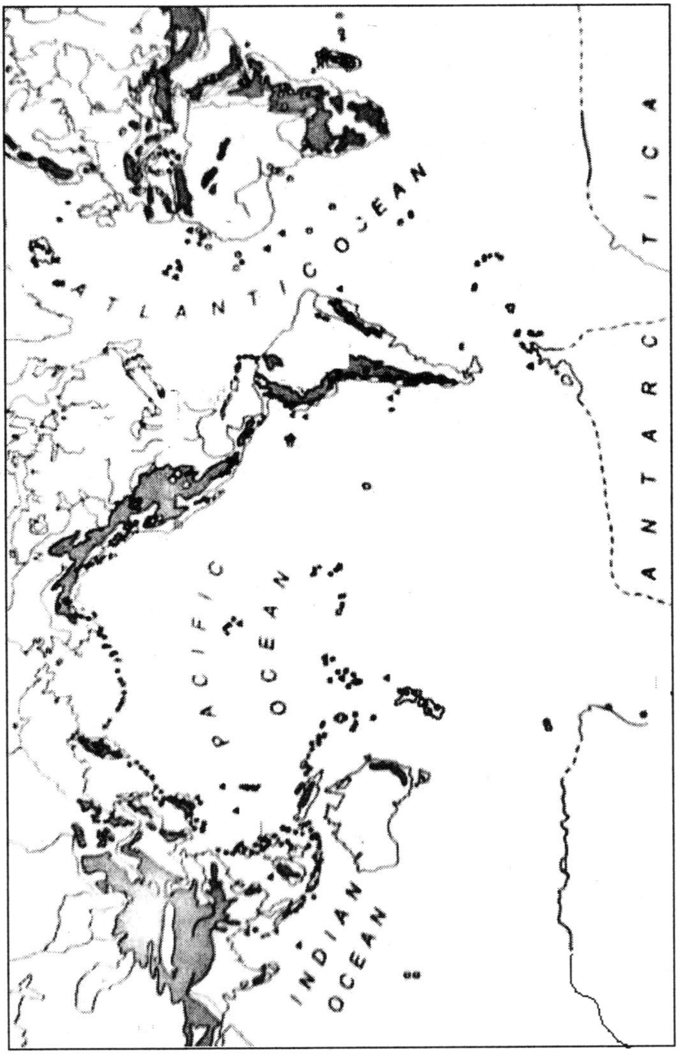

Source: Singh, Savindra: *Physical Geography* (2010), Prayag Pustak Bhawan, Allahabad, p. 89.

Fig. 44: World distribution of Volcanoes.

1. Circum-Pacific belt: This belt is also known as the volcanic zone of the convergent oceanic plate margins which includes the volcanoes of the eastern and western coastal areas of the Pacific Ocean (or the western coastal areas of Asia). This belt is also popularly known as the "fire ring of the Pacific" or "ring of fire" because of intense vulcanicity. In this belt, volcanic eruptions are mainly associated due to collision of American and Pacific plate because of the subduction of Pacific plate below the Asiatic plate. About two-thirds of the world's volcanoes occur in this belt. This belt starts from Eremus mountain or Antarctica and runs northward through Andes and Rockie mountain of south and north Americas to reach Alaska from where this belt runs towards eastern Asiatic coast to include the volcanoes of island arcs and festoons (e.g. Sakhalin, Kamchatka, Japan, Philippines etc). This belt ultimately merges with the mid-continental belt in the east Indies. Most of high volcanic cones and volcanic mountains are found in this belt. Most of the volcanoes found in chain, e.g. the volcanoes of the Aleutian Island, Hawaii Island, Japan etc.

The boundaries of the Andes contain several hundred volcanoes, many of which are active now. About 22 volcanic mountains are found in group in Ecuador wherein the height of 15 volcanic mountains is more than 4560 m. above the sea level. Aconcagua, Tupungato, Cotopaxi are some high volcanic mountains in this region. In the Cordilleras of North America, there are about 12 volcanoes which are functioning currently. In the north of the Cordilleras, there is intensive volcanic activity. A chain of volcanoes nearly 3200 km in length can be treated through southern Alaska, the Alaska Peninsula and the Aleutian Islands one of the most active vulcanicity part of the world. In Kamchatka, there is concentration of about 129 volcanoes, 28 of which manifest contemporary activity. In Japanese Islands there are about 184 volcanoes, 55 of which were functioning since the historic period of time. Mt. Fuziyama, Bandai are famous volcanoes of Japan.

2. The Mid-Continental belt: This belt is known as the "volcanic Zones of Convergent continental plate margins". It includes the volcanoes of Alpine mountain range, Mediterranean Sea and also the volcanoes of fault zone of eastern Africa. According to plate tectonic theory, the major cause of volcanic eruption in this belt is the convergence and collision of Eurasian plates and African plates and Indian plates. The famous volcanoes of the Mediterranean Sea are such as Strombli, Visuvious, Etna etc. and the volcanoes of Aegean Sea are also included in this belt. To the east of this region, there are volcanoes of the Cancasus, Elburz Mountains and the volcanoes of Iran and Baluchistan. It may be pointed out that the main Alps and the Himalayas are entirely free from vulcanicity because of compact and thick crust formed due to intense pressure on the rocks of these mountains and also to the absence of vertical cracks.

In Africa, most of the volcanoes are found on the fault zone of eastern Africa. This zone is stretched over a territory of 3.5 thousand km from the mouth of the Zambezi river in the south to Somali in the north. The important volcanoes of this fault zone are Kilimanjaroo, Meru, Elgon, Birunga, Rungwe, etc., some active volcanoes are situated in the Red Sea also.

3. Mid-Atlantic belt: Most of the volcanoes of this belt occur along the mid-Atlantic ridge which represents the splitting zone of plates. This is the area from where two plates diverge in opposite directions (the American plate moves westward and the European plate moves eastward) and giving rise to fissure type of volcanoes. Due to divergence of plates there is constant upwelling of magmas along the mid-oceanic ridge and wherever the crust becomes thin and weak. Fissure flow of lava occurs because of creation of fracture. Some of the large Islands of the Atlantic Ocean are volcanic and most of them situated on the Mid-Atlantic Ridge. Jan Mayen Island in the north marks the starting point of volcanoes of this ridge. To the south of this, there is Island, which is the most active volcanic area on this ridge. The Laki fissure eruption in 1783 A.D. was so quick and enormous that 15 cubic kilometers of lava was poured out from 28 km long fissure. Recently, Hekla and Helgafell volcanoes erupted in the year 1974 and 1973 respectively. Other more active volcanic areas are Lesser Antilles southern Antilles, Azores, St. Helena etc.

4. Intra-Plate Volcanoes: Besides the three well defined zones of volcanoes as described above, scattered volcanoes are also found in the inner parts of the continents and oceans. These are the intra-plate volcanoes. The mechanism of the eruption of such volcanoes is not yet precisely known. The Pacific Ocean represents the largest lithospheric plate on the earth which is circumscribed by the 'Ring of Fire'. However, there are large numbers of intra-plate volcanoes which lie away from this ring. One branch of intra-plate volcanoes of pacific runs from Hawaii to Kamchetka.

Causes of volcano

Volcanic eruption mostly takes place in the weaker zones of the earth's surface. These weaker zones are attached with the destructive or convergent plate margins and the fracture zones of divergent plate margins. It also occurs along the divergent plate boundaries at the splitting zones of mid-Oceanic ridges. The mechanism of vulcanicity and volcanic eruption is associated with numbers of interlinked processes. These are (i) Increase of temperature with increasing depth as a result of heat generated by the disintegration of radioactive elements inside the earth, (ii) Splitting of plates and their diverse movement reduce the pressure overlying strata and thus lowered the melting point to enhance the formation of magma, (iii) percolation of rain water and melting water into the interior of the earth which in turn generates gases and vapour through heating process, (iv) Increasing volume of gases and vapour forced to ascend outside, and (v) Increasing amount of gases and vapour inside crust finally caused volcanic eruption and the intensity of eruption merely linked with the nature of crustal surface and the pressure induced by gases and vapour.

Volcanism and volcanic eruption can better be explained with the help of plate tectonics. Volcanic eruption and plate boundaries are closely associated. It is now well established that the plate boundaries and plate movement are the determiners of nature and intensity of volcanic eruption. The thermal convective current originated in the mantle beneath the crust creates tremendous pressure in the plate boundaries and thus cause the plates to move in opposite direction from each other. The splitting of the plates and their diverse movement creates fractures and faults which in turn releases the pressure in the underlying zone of mid-Oceanic ridge and also lowered the melting point. The materials of upper mantle are melted at a faster rate and started moving upward as magma under the increasing volume of gases and vapours. Increasing pressure due to the increasing volume of gases and vapour forced the materials to come out through the weaker zone developed as a result of plate movements. The materials so come out with the tremendous force usually cause fissure eruptions of volcanoes along the Oceanic ridge. The lavas thus come out cooled, solidified over the surface, and added to the trailing ends divergent plate boundaries.

This process of accumulation ultimately creates new basaltic crust. Sea floor spreading and divergence of plates in the Mid-Atlantic ocean give birth to the volcanic eruptions in Iceland and other islands located there inadvertantly, the mid-oceanic ridges, representing splitting zones, are associated with active volcano. As there is a constant supply of basaltic lavas from beneath the mid-oceanic ridges, therefore, most of the active volcanoes are attached with the mid-oceanic ridges. But most of the volcanoes away from the mid-oceanic ridges are inactive since supply of lavas decreases along with the increase of distance from the oceanic ridges.

Highly explosive types of volcanic eruptions are usually found in areas of destructive or convergent plate boundaries. When two convergent plates collide each other along the subduction zone, the subducted plate margin is started melting as it reaches a depth of nearly 100 km or more in the upper mantle. Magma thus formed is forced to ascend and resulted in the violent volcanic eruption on the surface of the earth. This type of volcanic eruption is primarily seen along the convergent plate boundaries. Volcanoes of the Mid-continental region and the Circum pacific belt are few examples of this type of volcano.

Consequences of Volcanic eruption

It is known to all that volcanic eruptions cause severe damage to both human life and property and also brought changes in the natural scenario of the region of its influence. Hot lavas and fall out of volcanic materials brought a total destruction of human structures such as road, railways, buildings, factories, airports, river dams and reservoirs. It also brought changes in the climatic condition and the natural settings of the region under areas of their influence. The consequences of volcanic eruption may be summarised as given below—

1. In absence of specific prediction in advance regarding probable volcanic eruption there will be tremendous losses to human lives. Sudden violent eruption through central vent will not give time to human beings to take any precautionary measures and in consequence the loss of human lives become severe. For example, the volcanic eruption in 1902 in the Martinique island of West Indies destroyed the town St. Pierre and killed a total of 28,000 inhabitants. Volcanic eruption accompanied by heavy

rainfall causes severe mudflow on the mountainous slope which causes death of human lives, Mudflow during volcanic eruption in Japan in the year 1919 killed a total of 5,500 people.

2. Infall of large amount of volcanic materials in the form of dust and ashes, fragmented materials, smokes etc. envelopes extensive ground surface and thus cause damages to crops, vegetation, buildings and the open water bodies. It also disrupts and diverts natural drainage system, spreads health hazards by emitting toxic gases during the eruption. It may even cause acid rains in areas of high humidity.
3. Sudden release of hot and liquid in large volume bury man made structures, causes death to both men and animals, destroy the forest areas. The great eruption of Mt. Loa in 1783 of Hawaii Island poured out huge volume of lavas which travelled a distance of 350 km from the volcanic vent and destroyed 15 agricultural farms along with the killing of nearly 24 per cent of the total population of the Island.
4. Normally earthquakes occur eventually before and after the volcanic eruptions which in turn destructive tsunamis in the sea. It develops most destructive and disastrous sea waves and causes death to human beings living particularly in the coastal belts. The 1883 Krakatoa's volcanic eruption developed disastrous tsunami which generated sea waves of 30 to 40 m. high and caused death to nearly 36,000 people of Java living in the coastal areas of the Island.
5. Changes in the climatic condition in and around the areas of volcanic eruption is more or less common. Increasing concentration of volcanic dust, ashes and smoke in the sky reduces the intensity of insolation reaching the earth's surface as they scatter and reflect back a considerable amount of incoming solar radiation. The loss of heat through long wave terrestrial radiation, however, continues and thus cause a sizeable reduction in surface temperature. Eruption of Krakatoa volcano in 1883 ejected dust, ashes and smokes covering nearly 20 cubic km up to a height of 23 km in the sky. It develops a thick dust veil in the atmosphere, which results in the reduction of solar radiation by 20 to 30 per cent.

6. Volcanic eruption also causes a total destruction of several ecosystems on the earth's surface. Scientists believed that volcanic eruption and fall of dusts, ashes fragmented materials cause mass extinction of enumerable animal species. Acid rains accompanied by volcanic eruptions also cause severe damage to both the plants and animals. Extinction of giant dinosaurs about 60 million years ago is also attached with the increasing volcanic activity during that period.

8
Geomorphic Processes

The geomorphic processes are the physical and chemical changes which caused modification of the earth's landform features. A geomorphic agency, on the other hand is a natural medium which has the ability to secure and transport materials of the earth surface. Wind, glacier, running water including concentrated and unconcentrated run-off, wave currents, tides, tsunami etc. are the important geomorphic agencies. All these agencies have the ability to remove materials from one part of the earth's surface and deposit them elsewhere by means of transport. Most of the geomorphic agencies are atmospheric phenomena and controlled by the gravitational force of the earth. W. Penck referred these geomorphic agencies as exogenetic force as all these agencies originate outside the earth's crust. On the other hand, geomorphic processes those originated within the earth's crust are designated as endogenetic forces. Earthquake, volcanism, diastrophism etc. are examples of endogenetic forces (The description of the mechanism of endogenetic processes and resultant landforms are discussed in the Chapter 4). The earth's surface of today are practically the product of the interplay of both exogenetic and endogenetic forces. Infall of meteorites also cause damages to the earth's surface, but this process is not included in the above-mentioned forces. The process can better be termed as extra-terrestrial process since it is typically a space born phenomenon. Geomorphic processes therefore, are grouped into two major categories, i.e. (i) Terrestrial process, and (ii) Extra-terrestrial process. Terrestrial process is primarily of two types, i.e. (i) Endogenetic process, and (ii) Exogenetic process.

Eventually, there arises a confusion regarding the common geomorphic processes. Some of the writers include weathering as a part of erosion. Weathering simply confines within the disintegration process of rocks and in no way is a part of erosion. Again, there is also a difference of opinion as to whether transportation is a part of erosion. Certainly, the transportation process as part of erosion extends the meaning of erosion too far to consider aggradations as a part of it.

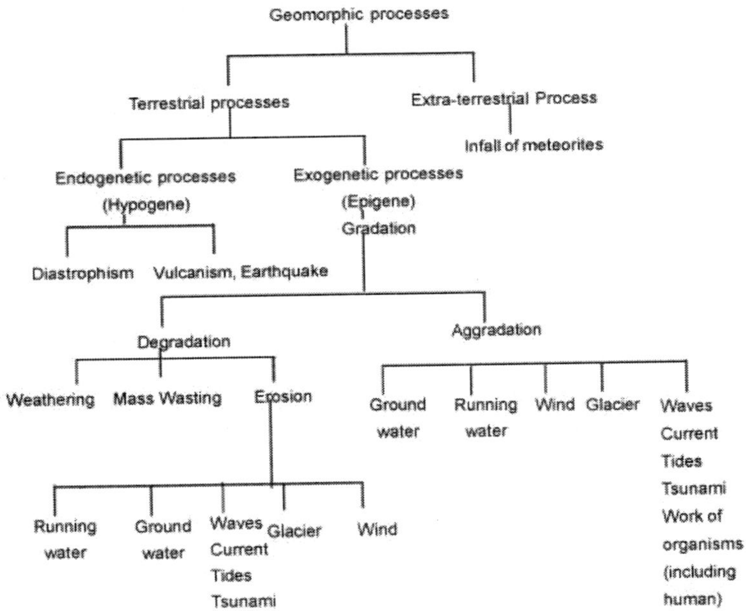

Fig. 45: An outline the Geomorphic processes

Chamberline and Salisbury (1904) very rationally forward the definition of the term gradation. Accordingly, gradation includes "all those processes which tend to bring the surface of lithosphere to a common level." Gradational processes belong to two types—those which level down, degradation and those which level up, gradations. The term denudation more often used as synonym of gradation by some of the geologists. The term **planation** is also used as alternative to erosion but gradation only implies erosion not the deposition.

Gradational processes

Degradation: Weathering, mass wasting and erosion are major degradational processes. Weathering defined as the disintegration and decomposition of rocks in place. It is primarily concerned with the processes, which act at or near the earth's surface and reduce solid rock masses by physical disintegration or by chemical decomposition to the elastic state and make the materials transport worthy. The weathered materials accumulated at the foot of the parent rocks are called regolith, which covers up the bedrock until displaced by transporting agencies.

Weathering

Weathering is mainly of two types (a) physical or mechanical weathering, and (b) chemical weathering. Along with physical and chemical processes of weathering, biological agents also actively take part in the process of disintegration and decomposition of rocks. At least four variable factors determine the type and rate of rock weathering. These are rock structure, climate topography and vegetative cover. Weathering does not make landform but only altered rocks or broken rocks. Weathering is therefore, studied comprehensively as a set of mineralogical or geochemical processes, because minerals are the chemical elements of which the landform are made. Every mineral has different chemical and physical responses to the prevailing environment. The works of exogenetic processes are essentially confined within a certain depth of the earth up to which exogenetic powered processes can operate. In an average of one kilometer, only the exogenetic processes are effective and its intensity is inversely proportional to depth.

Mechanical weathering or physical weathering

Mechanical fracturing of rocks is control by pre-existing cracks in or between mineral grains. Cracks in rocks are common and is formed as a result of thermal, mechanical and chemical stresses under varying atmospheric condition. Primarily five weathering processes are physical in nature and lead to fragmentation of rocks. Following Reiche (1950), an eminent geologist, the fragmentation of rocks may be designated as expansion resulting from unloading, crystal growth, thermal expansion, organic activity and colloid plucking.

Expansion resulting from unloading

One most important process of rock disruption related to physical weathering results from unloading of rocks. Unloading occurs when large rock mass are exposed through the erosional removal of overlying rocks. The rocks formed at great depth are in a compressed state as a result of pressure exerted by the overlying rocks. As they become exposed to the surface they expand slightly in volume. The co-efficient of elastic expansion of typical rock is approximately 0.1-0.8 per cent (Birch, 1966). Expansion accompanying the unloading causes overlying rock shells to break free from the parent rock mass. In massive rocks, fractures develop parallel to the land surface known as **sheeting**. Eventually, sometimes fractures in rocks developed thin onion-like layers or concentric rock layers. Such a process is known as **exfoliation.**

Sheeting is a near surface phenomenon and is mostly common in massive rocks like igneous intrusive, quartzite, and thickly bedded sand stones. Granitic monoliths in Yosemite National Park of California are an example of exfoliation dome. Exfoliation domes are produced as a result of sheeting of rock layers over the parent rocks. Bradley (1963) has described large scale exfoliation on numerous sandstone in the Colorado plateau.

Crystal growth

Crystal growth may result in rock fracturing. This includes not only the formation of ice crystals in rocks but also the formation of other crystals. Saline crystals are formed in dry climates as a result of the capillary action of water containing salts in solution. Certain minerals such as pyrite (FeS_2) in shale or slate and oxides are readily on exposure to oxygen bearing ground water or moist air. The newly formed iron oxides are of lower density but larger in volume. The increased volume of iron oxides exerts pressure in the rocks and split weaker rocks. Numerous hydrated salts that are water soluble are also effective to weathering processes. Ground water percolates upward into a rock and evaporates that left the salt precipitate in the rock cracks. Rainwater containing salt also left salt accumulates in the rocks. The force of crystallization of the precipitate salt is substantial. The formation of salt crystal in rock cracks or the openings is regarded as mechanical weathering

Formation of ice crystal is most effective as a weathering process when there caused repeated freeze and thaw. Expansion of rock cracks take place along with the freezing and then followed by contraction as a result of thawing. Repeated expansion and contraction of rock cracks leads to the fragmentation of rocks. This usually happens in the middle and high latitudes during early winter and late spring. Large daily range of temperature and frequent alteration of cyclonic and anticyclone weather is conducive to maximum freeze and thaw.

Thermal Expansion

Rock is naturally a poor conductor of heat and most of the rocks are composed of minerals having co-efficient of expansion. As a result of continuous heating and cooling, the minerals of the rocks response in a different way in the process of expansion and contraction powered by solar radiation. Most rock forming minerals expand when heated and contract when cooled. The alternate heating and cooling accompanied with expansion and contraction exerts a powerful disruptive force on the rocks. In the hot dry regions of the earth the difference between the day and night, temperature often exceeds 30°C. The extreme temperature at the daytime causes the thin upper surface layer of the rock to expand and to pull it away from the underlying cooler layer of rocks. This process is known as mass exfoliation. Minerals of varying quality response differently to the heating and cooling processes and thereby the rocks undergo granular exfoliation or disintegration. Laboratory experiments confirmed that under normal condition massive rocks have to be weakened first by chemical weathering before shells and flakes of rocks can be pulled away to open the way for fragmentation of rocks. Unequal thermal expansion of dominant minerals in the rocks is the controlling factor. For instance, the thermal volumetric expansion of quartz is three times higher than that of feldspar. Several minerals response differently to the heat and thus differ in expansion.

Organic activity

The wedging action of growing plant roots is often ascribed as mechanical weathering. The tree roots intrude into the bedding planes or joints of rocks for many meters. The growing roots exert pressure on the rock and force the cracks to enlarge. Roots follow the paths of least resistance and adjusted with the irregularities of a crack. Roots just follow the cracks opened by other processes. However, decaying vegetable matters and other plant debris keep rock surfaces wet and chemically active. Decomposed plant and animal matter may contribute to chemical weathering through the formation of carbonic and organic acids. Earthworms, ants, prairie dogs, and other burrowing animals open up varying means by bringing fresh rock material to the surface and make prone for chemical weathering. In addition, when trees uprooted or away in strong wind, their roots dislocate rocks in their attachment. Human activity in the name of development also accelerates the pace of weathering of rocks.

Colloid Plucking

Colloid Plucking is also considered as one of the weathering processes but nowhere it is found extensive and impressive to cause any form of substantial change in the rocks. It is possible that soil colloids have the power to loosen or pull off small bits of rock from the surfaces with which they come in contact. Colloid plucking may take place after each wetting and drying of soil colloids. No specific and distinctive evidence of colloid is get available. Frequent wetting and drying of soil colloids take place in the humid tropic which might able to cause changes in the rocks and helps to accelerate the pace of weathering.

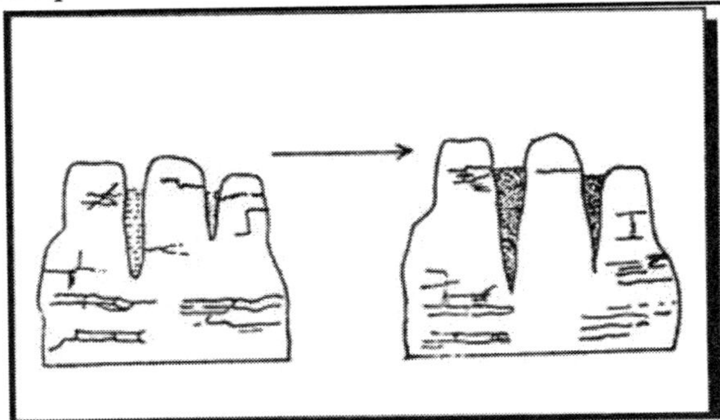

Fig. 46: Weathering of rocks by Frost

Fig. 47: Granular disintegration

Fig. 48: Exfoliation

Source: Khullar, D.R.: *Physical Geography* (2014), Kalyani Publishers, Ludhiana, p. 290.

Chemical Weathering

Contemplationally it can well be concluded that chemical weathering is more impressive and prevalent than other processes of weathering. Chemical weathering is pervasive and dominant particularly in the warm and humid climates of the equatorial, tropical and sub-tropical areas where heat and moisture are prevalent. Water being a reagent of solution is effective for chemical weathering. Water in contact of oxygen and carbon dioxide becomes an active reagent for chemical reaction. High concentration of oxygen, carbon di-oxide and water vapour in the lower most layer of atmosphere favours all kinds chemical reaction in the rocks of the earth's surface under influence which rocks are decomposed, dacayed and disintegrate into smaller particle size. Most chemical weathering results in : (a) An increase in size of rocks with resulting strains and stresses in the rocks, (b) Lowered density of minerals, (c) An increase in surface area between rock particles and the surrounding media, (d) Make materials more mobile, and (e) Produce more stable minerals. Among all these the important one is the increase in surface area interface which enhance the rate of reaction between rock material and the liquid as gaseous medium that comes into contact. (Jackson and Sherman, 1953). Chemical weathering is a process by which rocks are disintegrated and decomposed through chemical reaction ranging from simple to complex. Water vapor and water are essential component for chemical reactions in the rocks of the earth. Water when combined with atmospheric carbon-dioxide (CO_2), it becomes solvent to take part in chemical reaction. Oxidation, carbonation, hydration, solution, hydrolysis etc. are few important chemical reactions, which in turn lead to disintegration and decomposition of rocks.

1. Oxidation: Oxidation takes place when dissolved oxygen in water comes into the contact of rock minerals. Normally rainwater surface effects of oxidation are most apparent in rocks containing iron in the sulfide, carbonate and silicate forms, discoloration of minerals by the process of oxidation is readily noticeable. This process also affects minerals like amphiboles and pyroxenes. The oxidation of iron-bearing rocks reproduces rusts following the reactions:

$$Fe + H_2O + O_2 - Fe_2O_3 3H_2O \text{ (rust)}$$

Rusting of rocks, in turn, weakens them and amplifies the disintegration process of the rocks.

2. Carbonation: Atmospheric carbon-dioxide after mixing with water forms carbonic acid and become an effective re-agent of chemical reaction with rocks of the earth's surface. All types of carbonate rocks are susceptible to carbonic acid (H_2CO_3). Carbonic acid is formed when carbon-dioxide mixed with water ($CO_2 + H_2O = H_2CO_3$). Limestone dissolved in water and it forms calcium bicarbonate—$CaCO_3 + H_2CO_3 - Ca(HCO_3)_2$. The rainwater containing carbon-dioxide penetrates through the soil layer and come into contact of limestone. The solution of limestone produces varying types of landscapes are popularly known as **karst topography.**

3. Hydration: The process of hydration is related to the absorption of water by the rock minerals. Absorption of water by rocks causes change in their volume. The volume of hydrated rocks increases in size and may have an increase by two times than the original one on some occasion. Thus the increase in the volume of rocks produces stress and strain in the rocks surface and results in physical disintegration of the hydrated rocks. Exfoliation and granular disintegration of rocks usually caused as a result of hydration and prepare the mineral surfaces for further alteration by oxidation and carbonation. The following reaction represents the conversion of a hydrate to gypsum—

$$CaSO_4 + 2H_2O - CaSO_4 2H_2O$$

As such conversion of hematite to limonite also involves this hydration process.

$$2Fe_2O_3 + 3H_2O - 2Fe_2O_3 3H_2O$$

Both the reactions are exothermic and are reversible subject to the application of heat.

4. Hydrolysis: Hydrolysis process is associated with the formation of hydroxyl and ensures a chemical change in the rock's constituents. It is primarily concerned with the weathering of feldspars and micas. The chemical reaction can be presented as—

$$KAISi_3O_8 + HOH - HAISi_3O_8 + KOH$$

This alumino-silicic acid is unstable and produces colloidal silica, which forms clay minerals. The potassium hydroxide (KOH) will react in the presence of Carbon-dioxide and produce potassium carbonate. Potassium carbonate is soluble in water and is carried away as solution. Silicate minerals in general are mostly affected by hydrolysis process. Hydrolysis process, in fact, accelerates the pace of weathering in form of solution. Hydrolysis has resulted in the formation of magnesium hydrate, silicic acid and ferrous iron, which makes them soft to make way for fast weathering.

5. Solution: Solution is considered as the pre-condition for chemical decomposition and disintegration of rocks. Dissolution of mineral particles of rocks in addition of water or any often liquid medium is normally refers solution. Solution of rocks is related to some properties of rocks such as nature of rocks, solubility of rocks and the ratio between the solvent matter and the solid particles. Common salt is very easily dissolved in water. On the other hand, calcium carbonate, magnesium carbonate etc. are moderately soluble.

All Chemical weathering reaction involves water either as a reaction or as the carrier of the reaction products. As solutions in water are always part of all chemical weathering, therefore, it is seldom considered as a unique process of chemical weathering.

6. Chelation: The word 'Chelate' means 'Chawlike' refers to the tight chemical bonds where hydrocarbons impose metallic cautions. The role of iron in hemoglobin of blood to carry oxygen from the lung is a typical example. Plant leafs and other remains decay in the soil and add organic compounds that can chelate metallic cautions. Natural organic molecules in soil can remove metallic cautions directly from mineral grains by chelation. The mineral lattices are disrupted and fragmented. Vascular plants enhance chemical weathering by mutually reinforcing interactions among celation, hydrolysis and base- exchange reactions. The silica uptake by plants from basaltic rock is usually higher. The rate of silica withdrawal from igneous rocks is exclusive of any chemical weathering or leaching by percolating water. Even the no photosynthetic plants like Fungi have the ability to dissolve elements from minerals. During the incubation periods, the PH value normally is dropped and increases the acidity of soil and increases the solubility of minerals. Thus, fungal weathering is simply a biogenic process, which accelerates the pace of chemical weathering in acidified water. Humic acid activate chelation and help in the decomposition of silicate minerals.

Biological Weathering

Biological weathering is a type of weathering brought about by various activities of living organisms, which mainly include plants, animals and man. Biological weathering can be physical weatherings, chemical weathering or both. Along with other types of weatherings, biological weathering can contribute to the further degradation of rocks and rock particles by making them more susceptible to other environmental factors, whether be it biotic or abiotic factors. It may be pointed out that biological communities play an important role in breaking all types of rocks irrespective of climatic conditions. Biological weathering primarily is of the three types:

(a) Faunal weathering,
(b) Floral weathering, and
(c) Anthropogenic weathering.

(a) Faunal weathering: The burrowing animals, worms and other organisms help in gradual break down of rocks to a great extent. Burrowing animals such as gophers, dogs, foxes, rabbits, jackals, termites, rats, rodents, earthworms etc. dig out burrows and tunnels to make their shelter. By doing so they unconsolidate the geomaterial and break down the rocks to a great extent. Moreover, they bring new rock material from below over the surface and expose it to mechanical and chemical weathering. Small organisms play more important roles in rock and soil weathering. These organisms repeatedly mix up with the soil materials and thus always expose fresh materials to organic matter downward into the soil profiles and thus extend the weathering at greater depths which otherwise would have not been possible.

Earthworms in cool humid environments, termites in tropical climates and out in arid conditions remove the superficial overburden by their activities, forming impressive mounds of low dimensions. Termites in tropics pile up mud some up to 4 m. height. Termitaria are the evidences of soil weathering by termites. Worms consume large quantities of soil for the purpose of extracting food and the indigestible particles are passed out a worm-cast. It is believed that " in an average soil, there may be 1,50,000 worms to one acre, and in the course of a year they raise to 10 tonnes to 15 tonnes of finely comminuted materials from below to the surface. According to the estimate of Charles Darwin, the soil organisms bring nearly 25.4 thousand kilograms of soil per year from below to the surface in the English Garden. According burrow to about 1.5 m and pass 10 tons acre per year as the mean and 20 tons per acre per year to the maximum of the soil materials.

(b) Floral weathering: Weathering of rocks by plant is called floral weathering. It can take place in two ways viz. (i) physical weathering, and (ii) chemical weathering which is also called biochemical weathering.

(i) Floral physical weathering: In this type of weathering, plants exert physical force on rocks and weather them. Plants can grow anywhere as long as there is water. Roots of trees or plants in general can biologically weather rocks by penetrate into the cracks and fractures of rocks and soil. Cracks are widened by root penetration and the consequent root pressure leads to large scale weathering of the sub-surface rocks.

(ii) Floral chemical weathering: Besides their mechanical weathering, plants also act as an important agent of chemical weathering. Plants of all kinds, including fungi, algae, mosses, lichens etc. act as agents of chemical weathering because they abstract certain elements from rocks and librate others. When the roots of plants grow deeper into the soil, they tend to create cracks and crevices in marbles and limestones by producing certain acids that can eventually degrade them. When plants die, these takes place a decomposition of organic mass or humus which produces carbon di-oxide (CO_2). When this carbon di-oxide (CO_2), combined with water (H_2O), produces weak carbonic acid which can degrade the surfaces of rocks and rock particles. The minerals in rocks are librated when a fungus releases chemicals that can break them down. Such minerals are then consumed by the algae, further causing the wearing and development of cracks and gaps on the rock. As a result, cracked rocks become more prone to disintegration.

It may also be mentioned that apart from the weathering effect of vegetation on the rocks and soils, vegetation also protects the rocks and soils from weathering processes.

(c) Anthropogenic weathering: In recent times, man has become the most powerful agent of weathering with the development of science and technology. Man being a biological agent accelerates and decelerates the natural rocks of weathering by many ways. Mining, blasting of hills and ridges for construction of roads and dams and other developmental activities, quarrying etc. result in such a fast rate of disintegration of rocks. Man also accelerates the rate of weathering in hill-slopes by modifying the ground surface through deforestation and indiscriminate cultivation.

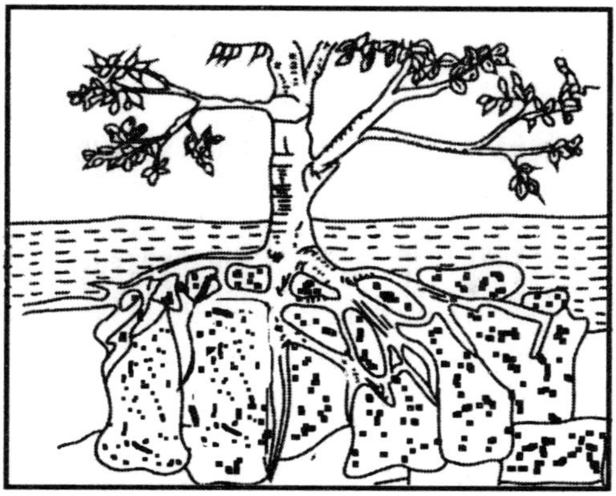

Source: Khullar, D.R.: *Physical Geography* (2014), Kalyani Publishers, Ludhiana, p. 295.

Fig. 49: Weathering by tree roots.

Mass wasting

Mass movement or mass translocation of rock wastes is simply known as mass wasting. The down slope movement of rock wastes in mass is accelerated by preened of water, ice and air under directional force of gravitation. The collective term for gravitational movement of weathered rock wastes is mass wasting. It indicates that gravitational force is the sole major force and there is no role of any other transporting agencies such as wind, running water, glacier, or molten lava. Water, although is excluded from the process of mass wasting but it plays an important role in mass wasting by over steepening the slopes through surface erosion of their bases. This activity of water is called lubrication of rock and weathered rock debris. Sharpe (1938) recognized four major types of mass wasting, which are designated as (1) Slow flowage, (2) Rapid flowage, (3) Landslides, and (4) Subsidence. Along with these major types, various sub-types are also recognized. A brief character of the various types of mass wasting is attached below:

1. Slow flowage type of mass wasting

(a) Creep: Gradual slope movements of soil and rock debris are usually visible except through careful observation.

(i) Soil creep : Down slope movement of soil.

(ii) Talus creep: Down slope movement of screed or small loose stones.

(iii) Rock creep: Down slope movement of individual rock blocks.
(iv) Rock glacier creep: Down slope movement of rock waste because of glacial activity.
(v) Solifluction: Rock debris saturated with water slowly flows down the slope following definite channels.

2. Rapid flowage type of mass wasting

(i) Earthflow: Downward movement along low gradient hillsides of water saturated clays earth material.
(ii) Mudflow: Movement of water saturated rock debris through definite channels.
(iii) Debris avalanche: Rapid flow of rock debris along a steep slope.

3. Landslide

A perceptible rapid movement of relatively dry masses of debris composed of earth materials along a steep slope:

(i) Slump: Downward slipping of several rock debris usually with a backward rotation along a hill slope.
(ii) Debris slide: The rapid sliding of unconsolidated rock debris without backward rotation.
(iii) Debris fall: Free fall of earth debris along an almost vertical slope.
(iv) Rock slide: The direct falling of individual rock masses down the bedding, joint or fault surfaces.
(v) Rock fall : Direct falling of rock blocks over any steep slope.
(vi) Subsidence: Downward displacement of surfacial materials without any horizontal displacement.

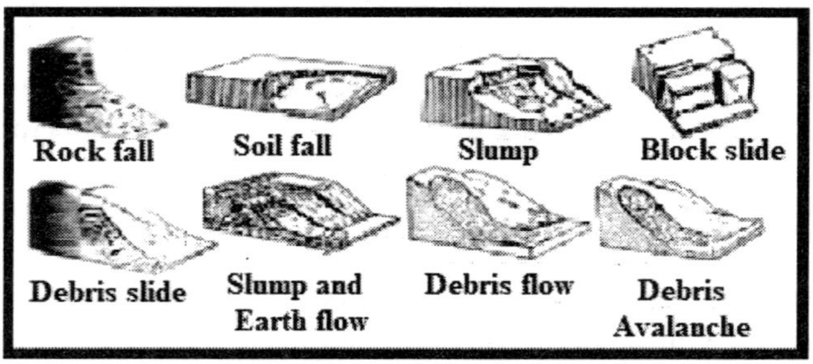

Source: Khullar, D.R.: *Physical Geography* (2014), Kalyani Publishers, Ludhiana, pp. 302-303.

Fig. 50

Various factors like lithologic factor, stratigraphic factor, structural factor, topographic factor and climatic factor play a vital role in the process of mass wasting. These factors aided by the processes like removal of support base through natural or artificial means, over-steepening of slopes by running water and other weathering processes, and overloading of rock materials through water saturation accelerate the pace of mass wasting.

Erosion

Erosion is primarily concerned with the removal of earth's material by various mobile agencies namely running water, wind, glacier, ocean waves etc. Among all these, water is the most important as it is effective and widespread as the normal means of land sculpture over the extensive areas of land surface of the earth. Sporadic rains occur in the arid region, where work of wind is predominant, plays a very effective role in the process of erosion. The various agents of erosion accomplish erosion by distinctive processes inherently related to the physical character of the agent.

The major landform features those developed as a result of tectonic activity are substantially modified by the persistent attack of the agents of weathering and erosion. Primarily, the process of erosion consists of four aspects. These are (i) The acquisition of loose material, (ii) the wearing a way of bed rock through impact of rock material in movement, (iii) The mutual wear of rock particles by friction while in transit, and (iv) transportation of the acquired rock debris. Under certain condition solution becomes a significant aspect of erosion.

It is common that the proportion of surface and subsurface water that flows to the streams varies significantly with climate, slope, vegetation, soil type, bedrock etc. Estimation made on the flow of water ensured that one-eighth of the total annual run-off the hydrologic cycle goes to the sea directly over the land surface and the remaining seven-eighth part of water goes underground. Both the overland and underground flow of water changes chemical character of some rocks. The infiltration capacity of soil depends on the duration and intensity of precipitation, humidity of soil, vegetation, soil texture and gradient of the surface. Saturated soil surface makes way for overland flow of water and loosened soil particles are wasted away by the running water. Surface run-off begin in the sun-dried and clay rich soil surfaces as soon as the top soil layer is saturated. Areas having dense vegetative cover slow down the run-off as well as the erosive capacity of the running water.

Sheet flow or overland flow runs only during rainstorm and the flow may be laminar or turbulent. Soil erosion and transportation of soil particles depend on the impact of raindrops on the soil surface. Surface run-off usually flow for a short time period as sheet flow. Surface tension and internal friction in flowing water soon develops ripples where turbulent energy is concentrated and thus generates ephemeral streams, which are called rills. After rain, the water soaks in the channel bed and thus intensifies the soil erosion. On the other hand, a gully is a steep sided ephemeral stream valley in which headword elongation is a common phenomenon. Gullies grow headward, capture adjacent rills, and thus develop a network of tributaries. Gullies eroded headward parts of its course and produce a water gathering wash-slope. Gullies are attributed to be the first step in the fluvial dissection of landscapes (Bloom, 1979).

Diastrophism

Endogenetic or hypogene processes usually originate at certain depth within the earth's crust. Diastrophism and volcanism are two major types of this endogentic process. The impact of this process is widespread as it builds up portions of the earth's surface. Diastrophic process is usually of two types (i) orogenic (mountain building with deformation, and (ii) epeirogenic (regional uplift without much deformation)

Mountain building activities took place on the earth at different geologic past and are episodic in character. It is surmise that between orogenies there persist a long duration of relative stability in the earth's crust and subject to slow epeirogenic uplift or subsidence of the crust. However, there persist some controversies regarding the nature of occurrence of both orogenic and epeirogenic movement. Certainly, such contrasting views regarding the nature of diastrophism will profoundly differ in certain geomorphic interpretation and analysis.

Volcanism

Volcanism is associated with the processes of outward movement of molten rock or magma from the interior of the earth. Molten rock or magma may be extruded through centralized vents called volcano or through extended and dispersed openings or fissures as mass eruption, which is termed as **Fissure eruption**. Some of the geologists define volcanoes as the vents or fissures from which hot and molten subsurface materials are transferred to above ground level. It is generally conceived that volcanoes are characterized by eruptive activity. Hot gases, liquids, molten rock shattered rocks etc. are forcibly or rapidly ejected from the openings in the earth's surface. On the basis of activity in the historic part volcanoes are classified as (i) active, (ii) dormant, and (iii) extinct.

Most of the geomorphic features are volcanic origin. It is estimated that nearly 2 million square kilometer of land surface have been covered to a depth of 500 -3000 meter. By lava flows in recent geologic time. (Rittman, 1962, xiii). Out of the total active volcanoes, about 62 per cent are in the circum pacific "ring of fire", another 22 per cent are in Indonesia, 10 per cent are in Atlantic and the remaining is in Africa, Mediterranean and Middle-East and in the mid-ocean islands.

Geochronometry, a new technique draws interest of geomorphologists as it permits precise dating of eruptive rocks. It is therefore, now become feasible to measure the volume of rock eroded from a dated volcanic landscape and provide meaningful means to measure the absolute rate of landscape modifications. It becomes feasible for geomorphologists to study contrasting erosional forms and rates in any climatic regions having volcanic landforms.

Infall of meteorites

Landform developed as a result of infall of meteorites is uncommon. Most unusual landforms on the earth's surface are those that originate from the impact of meteorites. Landform develops by the impact of meteorite are rather rare but approximately fifty well-substantiated meteorite craters are identified over the earth's surface. The craters turned into lakes or ponds and become the accelerating agents of degradation processes. All the meteorites craters are produced by extra terrestrial agents but the earth's gravitational force is responsible for the infall of meteorites.

9
Cycle of Erosion
Davis and Penck

All landscape is merely products of geomorphic processes ranging from endogenetic to exogenetic. A landscape has a specific developmental history where it faces a series of gradual changes. The primary or the initial landform passes through a series of sequential form and attained at the final or ultimate form. In this way, all landscapes have to face many successive stages of development, which can be termed as youth, maturity and old age. The evolutional process of a landscape passes through a definite course of development, which can be envisaged as a cycle. As one simple instance, a newly uplifted sea floor may possess no relief and almost featureless surface. It may develop to a maximum as the valleys are depending. Its slow but inevitable obliteration as the ridges are lowered by erosional process with the development of another surface of low relief or no relief. The recognition of the possibility of such eventual planation is considered as the most fundamental principle of cycle of erosion.

It is obvious that throughout most part of the world, run-off waters are the dominant geomorphic agency except for areas covered by glaciers. In the deserts even the impact of rain, although not occurred frequently, is pervasive in producing various landform. During Pleistocene period areas veneered with glacial deposits having unique topographic features, streams and sheet wash have already brought significant change of the landscape and are rapidly changing it. The prime concern at this context is no other than the erosion under the control of normal river action. The concept of the cycle of erosion may evidently be extended to cover the cases of arid and glacial erosion.

Eminent American geomorphologist W.M. Davis (1850 -1934) was the propounder of the concept of the cycle of erosion. This concept was formally put forwarded in the year 1899 and introduced as geographical cycle. Through this concept, Davis advocated that a landscape have a definite lifecycle where continuous changes in form and structure are brought by the processes of land sculpture. The changes thus occurred follow a definite sequence, which can be identified as initial form, sequential form and ultimate form on the basis of the changing surfacial character under the impact of the operating agencies. Thus, the sequence of the evolution of landscape passes through some successive stages, which can be envisaged as cycle. The cycle concept is primarily concerned with the process starting with the evolution of a new landscape with no relief at its initial stage and it passes through a sequence of change under varying geomorphic processes and again attain at a stage of no relief similar to the initial stage with some identical features.

Structure, process and stage

W.M. Davis advocated that landscape is a function of structure, process and stage. These three factors combined act in the process of landform evolution subject to no intervention of any deformities resulted from endogenetic forces like earthquake, vulcanicity etc. The term structure not only includes the regional geologic structure, folds, faults etc. but also includes the nature of rocks, their physical and chemical properties, their permeability and solubility and nature of joints, cracks, unconformities etc. A distinctive knowledge of geologic structure is essential to know the effectiveness of the varying geomorphic processes on the landscape. A rock may be 'hard or resistant' or 'weak or non-resistant' to geomorphic processes. Rocks are under continuous attack of both physical and chemical processes. A rock may be resistant to one process and non- resistant to another and under varying climatic conditions they may response differently. For instance, there may be minimum run-off and minimum erosion by running water in a surface composed of permeable rocks. Soft shales are eroded more rapidly than massive land quartzite rocks. A rock bed composed of limestone or gypsum will undergo solution by underground water. As a rule, the influence of rocks on landscape is quite pervasive. The structure of all landscape is older than the landscapes under operation of gradational agencies and thus the gradational processes give birth to a new sequential landform as time passes.

Process is dynamic in nature and brings about changes in the whole configuration of the landforms. Processes are grouped into two principal types, which are termed as endogenetic or internal, and the exogenetic or external. Marked irregularities on the surface of the earth, namely mountains, hills, plateaus, upland surfaces etc. are prime creation of endogenetic processes, which expose themselves through diastrophism, vulcancity etc. As soon as the features exposed to the outer margin of the earth's surface they come under attack of exogenetic forces or external processes which tend to level down the earth's surface. Running water, underground water, glacier, wind sea waves etc. are chief agents of external processes. Land uplifted by the endogenetic forces are levelled down by the external processes and the materials eroded by sub-aerial erosion are transported and deposited them in most of the low-lying areas. As a result of the continuous contradiction between the two opposite forces a landform feature comes into existence over the earth's surface. The landscape, which are once uplifted by endogenetic forces are level down by exogenetic forces and transformed those into a featureless level surface. Processes are important dynamic component in the evolution of landscape and following the work of processes varying landforms are developed as time passes, as a particular process give rise to distinctive landforms. For example, flood plains, alluvial fan, deltas etc. are resultant features of river erosion; caverns, underground streams, limestone caves are product of underground water; loess, sanddunes, Inselbergs, Barchans etc. are features developed by the action of arid erosion while moraines, drumlins etc. are produced only by glacial erosion. As the action of running water almost all over the earth's surface is pervasive and wide-spread. W.M. Davis, therefore constructs the concept of cycle of erosion based on the action of running water.

As the agents of erosion started working upon a land surface, it normally develops characteristically different landform as time passes. There is a gradual and systematic evolution of landscape. The sequence of evolution is divided into three stages namely youth, maturity and old stage. W.M. Davis introduced these three sequences of development in his concept, **the cycle of erosion**. Different processes give rise to different landscape and based on the characters of the evolution of landscape, it is quite possible to differentiate the stages of evolution of the landscape. There is a close relationship between the character of landforms and the stage of development. Time is a prime factor, which ultimately determines the stage that a landform attained. It would be misleading that the three stages have equal lifetime. Time factor is dependent on the nature of the structure and the character of the processes active on the landscape. The concept of cycle of erosion forwarded by Davis pertinently elaborated well regarding the sequence of stages a particular landform attained in its life span of evolution. Accordingly, a newly uplifted landmass possesses no relief or absents from having any intensive diversity of relief. As time passes, running water under humid temperate conditions develops river system and gives birth to active erosional activity on the landmass. Down-cutting in the river bed gradually increases the relief and at times will reach maximum and simultaneously slope wash, gulley erosion, rill cutting etc. processes engaged in removing materials from the hill slope will gradually reduce the relief. As a result, at times, the entire landscape turns into an almost plain surface with minimum or no relief. Based on their character of evolution of the landscape Davis indentified these distinctive stages in the process of the evolution cycle, i.e. youth stage, mature stage and old stage. The varying type of geomorphic processes working upon different landforms leave their distinctive imprint on the surface. The youth or initial stage attained at its final stage, which is termed as old stage by Davis through a sequence of development, and produce characteristic features at different stages and the distinction among various stages are apparently distinguishable. It is quite inferential that explication of the stages a landform attained on the basis of all distinctive characteristics will certainly enhance the perceivability of the geomorphic processes.

Cycle of Erosion

The propounder of the concept of cycle of erosion was eminent American geomorphologist William Morris Davis (1850-1934). It was put forwarded in the year 1899 in a well illustrated form which he called **geographical cycle**. All landscape posses its own specific life history during which it exhibit a series of gradual changes, whereby the initial landform changes through a series of sequential forms and ultimately it attains at last stage, the old stage or ultimate stage. Accordingly, the entire process can be broadly grouped into three major stages namely youth, mature and old stage. The evolution of landscape thus visualize as a cycle which encompasses definite course of development.

Comprehensively the concept of cycle of erosion can be illustrated with an assumption that a newly uplifted sea floor comes under the operation of erosional processes as it exposes to the atmosphere, the generator of exogenetic processes. The appearance of the uplifted sea floor is almost featureless at its initial stage. Rainwater it receives started accumulating in the areas comparatively low than other parts and will start flowing out as it over topped. The run-off obviously will follow the slope of the surface and remove materials its path of flow. In consequence, a flow line develops over the surface and thus gives birth to a river valley. The relief will grow to a maximum as the valleys are deepened. Consequently, the ridges are lowered by the continuous and inevitable obliteration of the surface materials with the ultimate production of a surface of low relief. In fact, it is the fundamental principle of the concept of cycle of erosion. It is obviously the prime importance to recognize the varying stages of evolution attained in landscape development.

Erosion under the control of normal river action is the basic concern of the cycle of normal erosion. The conception of the cycle may evidently be extended in some other measures to encircle the cases of arid, glacial and marine erosion also. It is essential to note that the structure of a landscape is a most potent factor in determining the form. It is quite imperative that the initial surface, the starting point of a cycle of erosion may be very varied in forms and structures. It is normally assumed that a landform newly uplifted from the sea floor possesses minimum or no relief. It is also assumed that the uplift was so rapid that there is no scope for any significant amount of erosion at the time of its progress. This newly uplifted surface will have an initial relief produced during the uplift. The linear flow of water follows the natural gradient of the surface, which turns, into a permanent flow line to give birth the stream. The courses of the stream thus developed will primarily be consequent upon the broader, original slope of the surface. Such streams are called **consequent streams** and concomitantly produce consequent valleys by the process of excavation in the early stage of the cycle. It left clues to trace the evolution of the landscape as a whole.

At the initial stage, the interstream areas or the interfluves will be broad flat-topped ridges and possesses extensive tracts of initial surface.

The interfluves become narrow with the progress of valley widening. Valley widening and valley deepening is interrelated. Valley widening will stop as the down cutting has ceased and as a result, the traces of initial surface are totally wiped out from the surface. It indicates the beginning of mature stage, which is characteristically distinct from youthful dissection. In the youth stage, tributary development will have been proceeding and the original interfluves are removed and subdivided. Tributaries develop to total maturity and relief become maximum height. The quintessence of the Characteristic features of the youth stage can be summarized as:

1. There will be a few consequent trunk streams but few large tributaries. Numerous short tributaries and gullies will be extending themselves by headward erosion and developing valley system.
2. Valleys will have V-shaped cross profile.
3. Except along trunk stream, there will be no valley development and valley sides will rise from the stream's edges.
4. Interstream tracts are extensive and poorly drained. Lakes and swamps exist.
5. Waterfalls and rapids exist where the stream cross beds of hard rock.
6. Stream divides will be broad and poorly defined.
7. Stream meandering may exist but they are incised below the upland surface.
8. Rills and gullies lengthen their longitudinal profile through headward erosion.
9. Vertical erosion is more prominent than the lateral erosion until the late youth stage.

In the mature stage, the interfluves take the shape of sharp-erested ridges developed by the intersection of adjacent valley sides. The progress of valley widening narrows the interfluves and eventually destroys the last surviving traces of initial surface. Tributary development proceeds in the youth stage and gradually the original interfluves have been cut into and sub divided. The tributary development depends upon the intensity of rainfall and perviousness of the rocks. The drainage networks develop fully and massive erosion takes place in the valley region. The mature landscape is considerably affected by quantity and the arrangement of the disintegrated rock waste, which covers a considerable portion of the slope. Residual deposits may accumulate on the flat tops of the interfluves but the valley sides tend to be broken by base rock outcrops. At this stage, there will also be an extension and evolution of drainage network with the development of numerous tributaries, river piracy, fluctuation of divides, flood plain widening etc. The river usually takes time to attain the graded condition as the grading of the head streams take a much longer time. The work of grading of the streams continues for a longer period as the grade is not completed in the entire river system at the same time that the work of grading of the stream continues starting from mature stage to the last of old or ultimate stage. At the same time, the rivers develop extensive flood plain and meandering of the river courses.

In the youth, the slope of valley side is steep and as a result, rock wastes climb down easily. But in the later stage, where slopes are gentler, they become entirely covered with a layer of fine-grained waste. Under the influence of soil creep or solifluction these wastes slowly move downhill. The slope as such gradually tends to be graded and become almost similar to graded stream course. As a result, the sharp divides of early maturity eventually transform into convex form. Convexity of the ridge-tops will not be attained until active valley deepening has ceased. The ridge crest will remain sharp if the valley deepening is continued after the stage of mature dissection. Such sharp mature divides are seen as common feature of many mountains. At that condition, valley-deepening proceeds simultaneously with an equal lowering of the ridge crests. Crest-top and valley floor are linked by a graded slope in active motion. Thus, the maturity of a landscape proceeds to old age by insensible gradation process. The prime characteristic features of mature stage are summarized as:

1. Valleys are extended themselves and develop a well integrated drainage.
2. Stream courses adjust with the lithologic variation and developed longitudinal valleys.
3. Stream divides are sharp and ridge like resulting in a minimum of interstream uplands.
4. Lakes and swamps exist in youth are disappeared.

5. Flood plain tracts occupy a considerable portion of the valley floor.
6. Meanders are conspicuous and free to shift their position over flood plain and thus give birth to occasional ox-bow lakes.
7. The width of the valley floors does not exceed the width of the meander belt.
8. The minimum possible relief exists in this stage.
9. Slopes and valley sides occupy maximum of its topography.
10. Deposition of sediments on either side of the trunk stream leads to the former of natural levees.
11. Alluvial fans and alluvial cones develop in the foothill zones because of substantial decrease in channel gradient and the transporting capacity of the rivers.

Decrease in channel gradient, absence of valley deepening, decrease in the number of tributaries and flattening of the valley are characteristic features of a landscape attained in the old stage. Along with the trunk stream, the tributary streams also attain the base level of erosion and are graded hence down cutting or valley deepening is totally absent as it attains in the old stage. Lateral erosion, downwasting and weathering processes become more active and lowered the absolute altitude and the water divides of the surrounding areas. Reduction of height of the interstream areas and water divides convert the whole region into lowland. Because of low channel gradient, the transporting capacity of the streams becomes minimum and as a result, the riverbeds are becoming overloaded with eroded debris. Sedimentation in the riverbed become more active, occasionally sand dunes, and sand bars emerge at the riverbed. Low channel gradient and the extensive floodplains slow down the flow velocity and in consequence the main channel of the river is divided into numerous distributaries to make the river a braided in character. Bank deposition gives birth to the levees in both side of the river. As such, deltas are formed in the river mouth under congenial condition of the point where the river discharges its water and sediments.

The landscape as a whole transformed into extensive plain of undulating surface termed as peneplain. In form of exception, the hard resistance rocks, which are left untraded, may exist in the peneplain called **monadnock**. The prime characteristic features can be summarized as:

1. Tributaries to trunk stream are few in number than in maturity.
2. Valleys are broad and gently sloping both laterally and longitudinally.
3. Flood plains are well developed and stream flow is in broadly meandering courses.
4. Valley widths are greater that the width of the meander belt.
5. Interstream areas have been reduced in height and stream divides are not sharp.
6. Lakes swamp and marshy land presents in large in the flood plain zone.
7. Mass wasting and chemical denudation are more prominent over other processes.
8. Extensive areas are at or near the base level of erosion.

W.M. Davis forwarded this concept as an ideal model to represent normal cycle of erosion considering the running water as single predominant process in the evolution of landscape. Completing the three successive stages, i.e. youth, maturity and old, a landform may transform into a featureless peneplain subject to absence of any intervention in form of diastrophic movement. Eventually, standstill position for a longer duration of crustal stability is hardly possible as the earth is very much unstable. For instance, the plate tectonic movement is actively working where plates are always in motion and hence long period of crustal stability is quite impossible.

Graphical presentation of cycle of Erosion

The cycle of erosion may be expressed graphically to have an understanding the evolutional process of a landscape. Of the two lines used in the graph, the upper one dashed line represents the maximum relief while the lowermost continuous line represents the resultant surface produced as a result of erosion. The dotted line OA represents the average initial height of the higher parts of the landscape, while OB represents the average height of the lowermost part of the surface. The straight vertical distance AB represents the average initial relief of the surface. During the initial stages the valley bottoms are lowered at a faster rate while the ridge tops remained unaffected. The relief, therefore, increases, as the two graphs represent in the diagram. The relief reaches a maximum at the early part of maturity, when the last traces of initial surface disappear. As time passes, the ridges begin to lower in height faster than the valley bottoms are lowered. At this stage, vertical erosion is more pronounced and lowering of ridges take place at a faster rate. Gradually two curves come close to each other representing the diminishing relief.

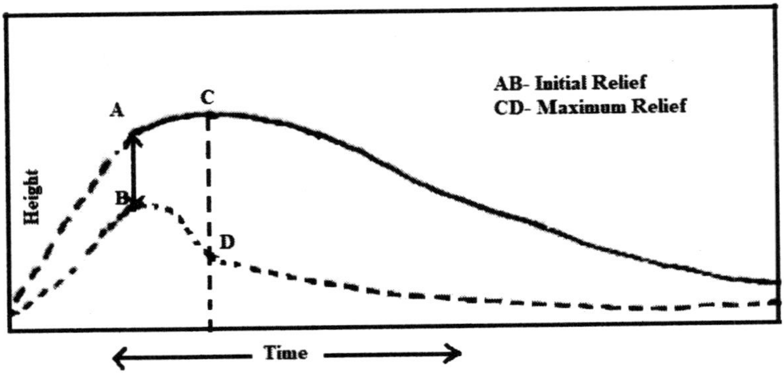

Fig. 50: Graphical representation of cycle of erosion (after Davis).

In the later stage of the cycle, the consequence of the two curves will slowly advance to merge. Since the processes immensely protracted therefore, intensity of erosion will decrease along with the reduction of height and steepness of the slope. The total convergence of the two will take a longer time and is not possible to specify the end on the graph. The Davisian concept of cycle of erosion as is represented on the graph is somehow meaningful to understand the whole conceptual base of the cycle of erosion.

Geomorphic Ideas of Walther Penck

The German scholar, Walther Penck offered a concept of geomorphology, which was a challenge to the normal cycle of erosion forwarded by W.M. Davis. Penck basically emphasized on the geological attribute in his concept on the landform evolution. Penck approaches on landform is primarily concerned with the viewpoint of a geologist who sought to find clues with an effective interpretation of the nature of the internal forces (endogenetic) which has a most penetrative effect on the topographic characteristics of any region. However, Penck did not completely repudiate the validity of geomorphic cycle forwarded by Davis. He viewed that unusual cyclical development of landform depend upon the stability of the earth's crust.

According to Penck, the characteristics of valley side slopes expose the relative importance of exogenetic and endogenetic processes. Penck recognized three types of slopes; there are uniform or straight, concave and convex. Uniform slopes developed under a constant intensity of erosion or **gleichförmige Entwicklung** (uniform development), concave slopes developed under decreasing intensity of erosion or **absteigende Entwicklung** (waxing or weak development), convex slopes developed under high intensity of **erosion or aufsteigende Entwicklung** (waxing development).

According to Penck, there are two types of slopes in the valley side. The upper slope, relatively steep are called bosche or gravity slope and the lower, relatively gentle are called haldenhang. Meyerhoff, with his observation on the characteristics of the slope called them gravity slope and wash slope respectively (Meyerhoff, 1940). The two slopes meet at a distinct angle. It depends upon the size of material rolled over them and the availability of water to carry the materials. Large sized material requires a steeper slope for its downhill movement. Generally, the haldenhang a wash slopes are steep in arid region than the humid areas. According to Penck, the gravity slope retreated upslope keeping unchanged the original angle of slope. This type of slope retreat is called parallel retreat of slope. As under conditions of waxing development, concave slopes develop with flatter surface which is almost similar to Davis' idea.

The successive stages in the evolution of relief are represented in the Fig. 51. As per Penck's idea, in stage-1, the flat interfluves are little eroded and increase in altitude by the amount of upheaval. Valley deepening does not keep pace with the upheaval and as a result both absolute altitude and relief increases. In stage-2, valley widening process makes sharp the interfluves and then interfluves are lowered as fast as the valleys are lowered. The altitude started increasing slowly by the amount of uplift minus the amount of valley deepening. In stage-3, valley deepening continues along with the upheaval of the landscape and deepens their valleys maintaining balance with the uplift. The ridges are lowered as fast as the valleys are deepended and in consequence, the relief remains constant. In stage-4, uplift completely stopped, but valley deepening continues for a specific time and altitude begin to decrease, but relief remains unchanged. At the end in stage 5, the process of valley deepening slackens and ridges become rounded which in turn result in the reduction of both relief and altitude.

The important aspect of Penck's concept is the balance between the rate of uplift and the rate of valley deepening in stage 3. Basically Penck opined that the uplift of a landform continues for a long period and as it exposed to the region of and as it exposed to the region of zone of exogenetic processes, the conflict between the uplift and degradation starts. Penck regarded the earth's surface as **"zone of conflict"** between external and internal forces and he believed that the landform reflected the relative intensity of both the endogenetic and exogenetic processes.

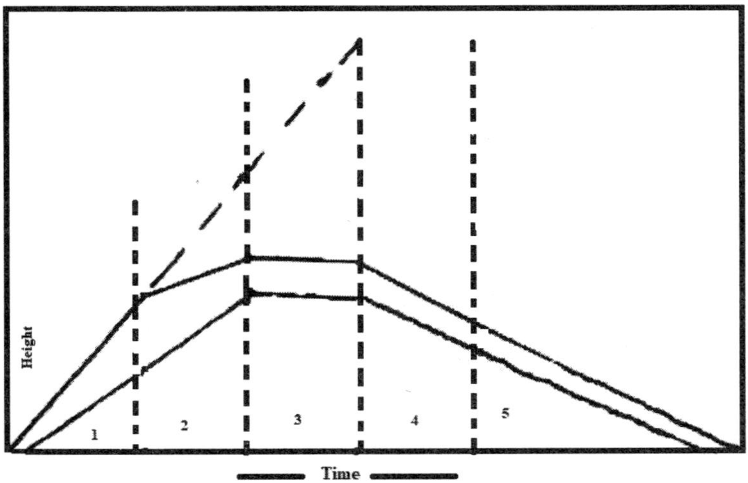

Fig.51: Penck's concept on cycle of erosion during long continued uplift. (Upper and lower line represents respectively the ridge crest and valley bottoms.)

The primary concept of Penck is indubitably a magnificient generalization as diastrophic conditions necessary for the progress of a cycle of erosion. Unlike Davis, the idea of Penck regarding the initial process of a cycle is associated with the slow uplift with respect to the rate of denudation. As a result, there will be no actual net increase in the relief. Such a condition would give rise to a featureless plain, which he called **primarumpf**. Slow initial uplift, irrespective of geologic structure landscape processes will certainly face degradation and would keep pace with uplift resulting in the formation of a primarumpf. The passage of a landscape through the successive stages of evolution will attain in its ultimate stage, which Penck recognized as endrumpf. Thus, the primarumpf is the initial stage in a period of diastrophism characterized by accelerated uplift, and the endrumpf is the ultimate stage of a period characterized by declining uplift.

10
Evolution of Landforms

Fluvial Landforms

Geomorphic features are product of geomorphic processes working under varying climatic conditions and on varying geologic structures. Landform produced under varying geomorphic processes reflects the inherited nature of the function of the processes and the structures of the parent body. Geomorphic processes operate upon any landscape leave their distinctive imprint on the surface and give birth to various landforms specific to their characteristics. The nature of denudation of geomorphic agencies such as running water, wind, glacier, ocean waves etc. normally develop varying types of landforms, which are distinctively different from one to another. Landform develops in consequence of erosion by running water are quite different from the landform develops under arid erosion. Similarly, landform in the glaciated areas, coastal region differs from each other in a significant manner. It is obviously necessary to have some discussion on the types and origin of landform develops under different conditions. A brief analysis is attached here for the learners to gain a conceptual idea on the various landform develops under varying conditions over the earth's surface.

Landform associated with fluvial erosion

Water always tends to flow through a straight course and its velocity is controlled by the surface gradient plus the availability of water. Usually, rivers are originated at sources like lakes, glaciers, springs etc. Some of the world's prominent rivers originate from lakes, namely, Lena and Yenisei from Lake Baikal, St. Lawrence from the great lakes and so on. Ganga from Gangotri, Yamuna from Yamunatri, Brahmaputra from Yangshi etc. are few examples of rivers originated from glacier. Most of the south Indian rivers are examples of rivers originated from springs. Rivers in its mountainous course have to flow along the steep slopes and thus velocity of water is higher and as it flows over the rocky surface, the intensity of corrosion or erosion is great. Because of extensive down cutting, the valleys are deep and comparatively narrow. River thus flowing over the varying nature of surface under varying climatic condition develops various types of landform. Such landform features are representation to the processes from where they develop. It is, therefore, quite speculative to the processes and the types of landform develops under various conditions in different parts of the earth surface. A brief description on the landform features develops under varying geomorphic processes is attached below to provide a general idea on the features.

Canyons: Rivers flowing over a mountainous or plateau region having less rainfall usually unable to erode the river valley sides by rain-wash. Only valley deepening proceeds with time. Thus, such rivers flow a long distance through a deep vertical sides. Such deep river

Fig. 52: Canyon

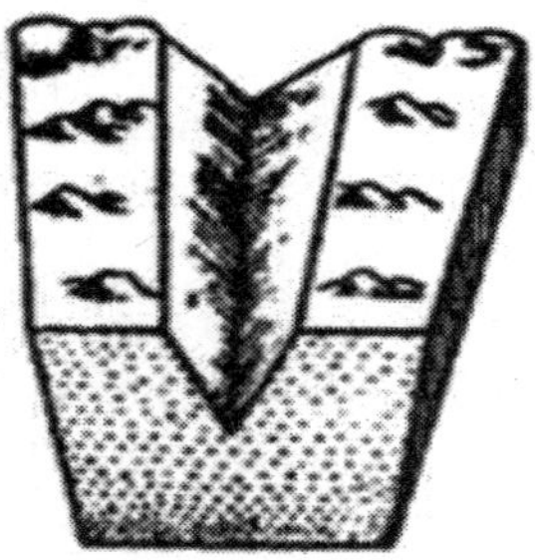

Fig. 53: Gorge

Source: Mitra, Kalpana, Sen, Jyotirmoy & Sengupta Purabi: *Physical Geography* (2014) Kalyani Publisher, Ludhiana, p. 308.

valleys with vertical steep sides are known as **Canyons**. The Grand Canyon on the river Colorado river of United States of America is 483 km. long with 1.5 km. depth and 12 km. width in average. Such

Canyons are also found in the river Hwang ho, Brahmaputra and Zambezi river also. Development of Canyon, of course, is associated with the rock type of the region over which it flows. Mountains or plateaus formed of soft and non-resistant rock to erosion is not congenial for the development of canyons.

Gorge: Gorges are normally formed in the upper course of a river where it flows through mountainous or hilly region. The velocity of the rivers is higher because of the steepness of the slope. Predominance of angular rocky structure in contact of high velocity of flow enhances the corrosive and erosive capacity, which in turn accelerates the pace of down-cutting of the river valley. The erosion is mostly vertical downwards and thus extends the pace of down-cutting of the river. The valley, as a result, become deep and narrow. The river tends to flow through a deep, narrow channel and in consequence it gives birth to the formation of a gorge. A deep and narrow river valley in the upstream areas develops a gorge, which is characteristically originated in the mountainous and hilly areas only. In the Himalayan mountain region most of the rivers namely Ganga, Brahmaputra, Indus etc. possess a number of such gorges in its course of flow. Perhaps, the gorges produced by the Changpoo in between the Gyala Peri and the Naamcha Barwa peak in Tibettan region is the deepest gorge in the world as it still has not been possible to measure the actual depth following a number of attempts made in the past. As such the gorge in the Indus near Gilgit measures a depth of nearly 17,000 feet.

V-Shaped valley: It has evidently been proved that valley deepening along with a parallel intervention of lateral erosion of the valley sides is necessary to produce V-shaped valleys in the upstream course of a river. In the mountainous course the river continues to cut its valley floor downwards and when such activities accompanied by a parallel process of valley widening under the impact of rainwater frost and other geomorphic agencies, the river valley gradually becomes **V-shaped**. In areas of hard and more resistant rock, the valley sides usually become steep and the shape of the V is seen elongated and narrow. Contrary to this in areas of soft rocks, the shape of V become wider. Rainfall, however, is a strong determinant of valley slope. The valley usually is found deep and narrow in areas where rainfall is scanty despite the presence of soft rocks in the surface. On the other hand, in areas of heavy rainfall the valley sides are found gentler and more open. The deep gorges and canyons developed in the salt range and Baluchistan are primarily due to the arid climate and contrary to this; the broader gently sloping valleys in the Eastern Himalaya and Nepal are product of heavy rainfall.

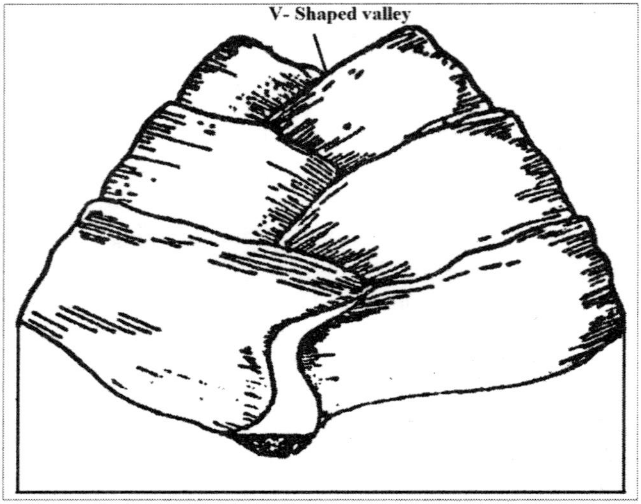

Source: Mitra, Kalpana, Sen, Jyotirmoy & Sengupta Purabi: *Physical Geography* (2014) Kalyani Publisher, Ludhiana, p. 310.

Fig. 54: V-Shaped Valley.

Pot-holes: The river course in the mountainous part is comparatively steep and as the water comes down along the river bed it altogether carries huge amount of rock debris. Depending on the buoyancy of the rock debris, some of the heavier rock debris is left on the bed of the river and in such condition, the river beds are littered with rock fragments. Subsequently, the rock fragments carried by the river waters face severe friction with the rock fragments lies at the river bed. Continuous collision among the rock fragments at the river bed accelerates the intensity of corrosion. Soft non-resistant rocks are more susceptible for corrosion. Where the river water flows over soft rocks and rocks having more joints and cracks are more susceptible for corrosion and small rock fragments are carry away by the river water. Under such condition hollows are formed at the river bed as a result of removal of rocks by friction and water current. These hollows become deeper and rounded on account of the rotary action of the rock fragments under the forces of moving water. Rounded depression thus formed in consequence of localized drilling by rocks fragment is called pot-holes. These are normally exposed in dry season as the river bed become shallow. Pot-hole is common feature formed in the Rocky River beds. The pot-holes found in Chottanagpur plateau are produced as a result of drilling in the river beds composed of varying types of rocks ranging from soft to hard resistant.

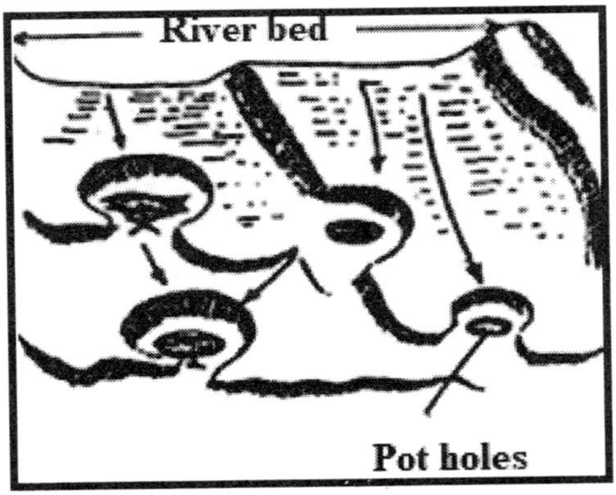

Source: Mitra, Kalpana, Sen, Jyotirmoy & Sengupta Purabi: *Physical Geography* (2014), Kalyani Publisher, Ludhiana, p. 308.

Fig. 55: Pot-holes.

Rapids : Rapids are swift flowing, turbulent flow of water usually developed in the upstream part of mountainous course. Streams in the mountainous region flow over rock beds of varying nature. As the stream line cross the rocky mountainous surface it has confront with rocks of both hard and soft in nature. Soft rocks are erodible and removed at a faster rate than the hard resistant rocks. In areas, where an outcrop of hard rocks dips (path of river flow) downstream and impedes the free flow of water and finally the accumulated water behind the hard rock is thrown rapidly downwards. These swift flowing, turbulent streamline with shallow waterbed are called **rapids**. Sudden steep inclination of river bed is certainly a precondition for the formation of rapids. In some areas, waterfalls may degenerate into rapids as the river becomes graded in consequence of the removal of the steepness of the rocky outcrop. Primarily features like rapids, waterfalls etc. are characteristic features of youth stage of a cycle of erosion.

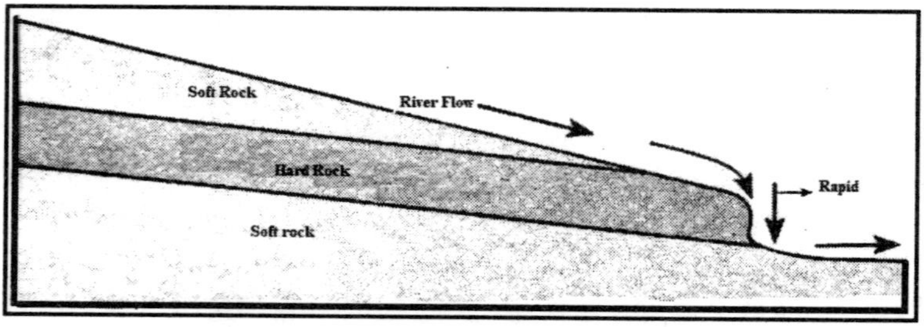

Fig. 56: Rapid.

Waterfalls: Waterfalls are formed when the water of river suddenly face a steep vertical slope and dropped down almost vertically

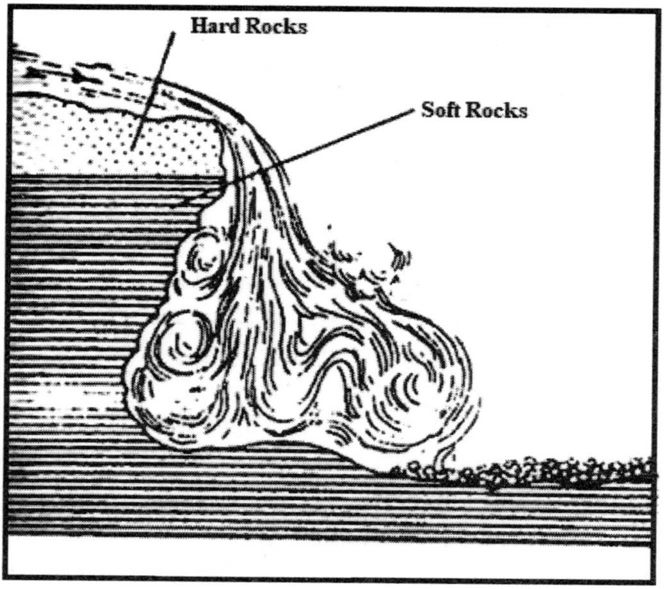

Source: Mitra, Kalpana, Sen, Jyotirmoy & Sengupta Purabi: *Physical Geography* (2014) Kalyani Publisher, Ludhiana, p. 311.

Fig. 57: Waterfall.

from a substantial height. The river water suddenly dropped down along a vertical slope usually gives birth to a **Waterfall**. Running water suddenly drops vertically downward is simply called waterfalls. The presence of a steep vertical slope along the course of a river is a must precondition for the formation of a waterfall. Such a slope may be formed because of differences in structure, differential erosion alternately over hard and soft rocks in the river's bed and tectonic uplift or subsidence of the landscape along the course of the river. Not all waterfalls are common in their type and structure. On the basis of structure and character, the waterfalls can be grouped into various types. There are (i) Cap-rock waterfalls, (ii) Vertical barrier waterfalls, (iii) Plateau falls, (iv) Fault plane falls, (v) Knick point falls, (vi) Hanging valley falls etc. A brief description on these types of waterfalls is attached below.

(i) Cap rock waterfalls: This type of waterfall is usually formed in areas where a layer of hard rock overlie horizontally above a layer of soft rock. The soft rocks are removed faster by erosion leaving the hard rock layer plunging over it. As the running water falls over the soft rock layer it makes a hole and gradually the hole is enlarged and increase in size and depth. As water continuously drops down over soft rocks, the pace of erosion also accelerate along with the gridding action of rock fragments. If, eventually the river has a large volume of waterfall and high descending velocity, the grinding of rocks at the base of the fall is more pronounced and results in fast under-cutting of the baral rocks. Removal of baral rocks also removes the support base of the overlying hard rock layers and in consequence, the pieces of hard rock started falling down at the base. Thus, headward recession of waterfall starts and proceeds further headward as time passes. Niagra falls in the St. Lawrence River of North America is an example of cap-rock falls. The cap rock of this fall is composed of hard and resistant dolomite and the underlying layer is consisting of soft shale. The quick removed of underlying soft shale layer gave birth to this fall.

(ii) Vertical barrier waterfalls: This type of fall is more or less common in areas of sedimentary rocks with intrusion of igneous rock in the form of granitic or basaltic down like a dyke. The soft rocks of sedimentary layer in the riverbed are lowered down at a faster rate, while the hard resistant rock of vertical dyke remains unchanged. As a result, the stream descends over the vertical dyke wall in form of a waterfall. Yellowstone river fall of U.S.A is an example of vertical barrier fall.

(iii) Knick-point falls: Rejuvenation occurred in the lower section of a river either because of changes in sea level or uplift of the land, there develops a sharp break in slope at the edge of rejuvenation head. The point of slope breaking is called **knick point**. If the rejuvenated part happens to be composed of hard rock, the river develops waterfalls at the point of slope break the knick point. These types of falls are commonly found in the Appalachian mountainous region. Such knick-point falls are also seen in the Chottanagpur plateau region of India. Hundru falls, Daram falls, Gersopa falls etc. are such falls found in India.

(iv) Plateau falls: Plateau falls are usually seen in areas where the rivers sharply descended to a plain from an uplifted plateau primarily composed of hard and resistant rocks. These falls are also subjected to headward retreat or recession. For example, the Congo River of Africa is descending down from the African plateau to the coastal plains of Atlantic Ocean give rise to the Livingstone waterfalls.

Fault plane falls: Sudden change in level as a result of uplift of the upper part of a river develops fall along the line of fault scarp. Such falls are found in most of the ancient uplifted plateaus. Subarnarekha River, Sivasundaram falls of Cauvery River, Behar Falls of Tamra River etc. are examples of fault plane falls.

Hanging valley falls: Intensive erosion in the riverbed of the master stream flowing through mountainous or plateau region is the congenial pre-condition for development of a Hanging valley falls. The erosion capacity or a down-cutting of the tributaries are as much intensive due to the less velocity and low volume of water and started discharging its water into the main river from a substantial height. In such a situation, the tributary streams appear to hang above the valley of the master stream. Such valleys are called Hanging valley and the water discharge point develops falls like features. Such type of falls are called **Hanging valley falls**. Rajrappa of Chottanagpur plateau along Damodar River is a good example of hanging valley falls. Hanging valley falls are generally seen along the sea coast at the juncture of land and sea along a mountainous line.

Waterfalls are features of youthful stage of a river. Attainment in graded condition gradually destroys the existence of waterfalls. Waterfalls are useful to generating hydro-electricity and it provides a scenic beauty to build a zone of tourism.

Characteristics of landforms in middle course

In the Middle area, the stream velocity becomes sluggish but water volume is higher than the mountainous course. As it started flowing through a plain course, therefore it could not carry the larger rock fragments and these are deposited elsewhere in the valley. After leaving, the mountainous course the decreasing velocity of stream flew started depositing the sediments it carries successively following the size and weight of the rock fragments and sands. Larger and heavier rock fragments are chiefly deposited at the foothill zones. The downhill flow takes a triangular shape in consequence of deposition of fragmented large size rock debris. The river descended downhill in several braided channels. The entire process of flow and deposition give birth to fan like structure at the base of mountain. In areas of steep slope, numbers of smaller streams emerge from the mountain and spectacularly the alluvial deposit is less, such structure is termed as **alluvial cone**. Thus, at the base if the mountain or at the foothill where the river enters into a plain course developed two types of features known as **alluvial fan and alluvial cone**. Sometimes, the alluvial fans developed at the foothill zone merge with each other resulting in the formation of extensive **alluvial plain**. Such features are known as **piedmont alluvial plain or Bajada.**

In the middle course of a river, obviously the transportation and depositional activity exceeds the erosional activity. The river normally follows a meandering course over the flat surface. As it flows over a flat surface, the flow velocity decreases and wherever it faces some relatively impediments, it turns round the impediments instead of eroding that part. It then flows through a course of curved path developing meanders. In every meander loop there is a concave side and subject to constant erosion, while the opposite side experience continuous deposition of sand and silt. The side faces continuous erosion become steep and is termed as **cut-off slope**. On the other hand, the side filled with the deposition of silt and sand become gently sloping and is called **slip-off slope**. The cut-off slope is primarily a product of erosional activity while the slip-off slope is characteristically depositional feature found in the middle course of a river. A complete loop of a meander is called **meander belt**.

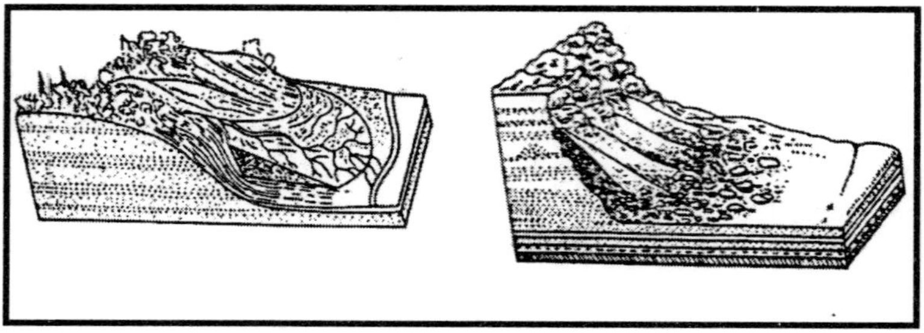

Source: Mitra, Kalpana, Sen, Jyotirmoy & Sengupta Purabi: *Physical Geography* (2014) Kalyani Publisher, Ludhiana, p. 312.

Fig. 58: Alluvial fan and Alluvial Cone

Ox-bow Lake: As Time passes, the curvature of the meander goes on increasing and finally there arrives a situation when the two ends of a meander meet each other because of continuous erosion in this part of the meander. The water of the meander started flowing through a straight course leaving the meander belt aside. The meander belt is completely isolated from the main channel and sand and silt are deposited near the threshold. The old stream course is thus physically separated and develops a zone if stagnant water bodies like a lake nearby the master stream. Such strongest water bodies are called **Ox-bow Lake** as it appears in the shape of an Ox-bow.

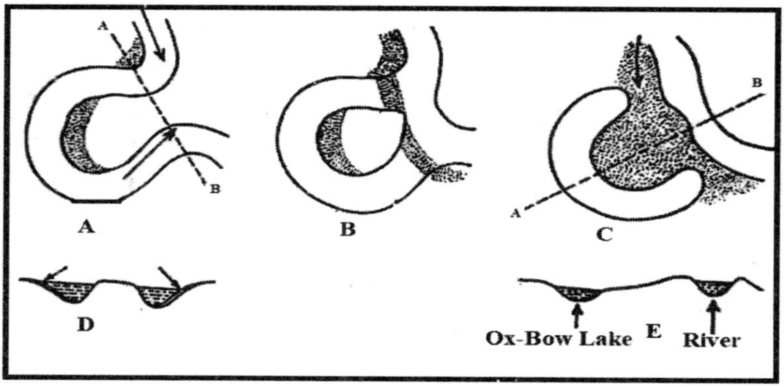

Source: Mitra, Kalpana, Sen, Jyotirmoy & Sengupta Purabi: *Physical Geography* (2014) Kalyani Publisher, Ludhiana, p. 314.

Fig. 59: Stages of Ox-Bow Lake Formation.

In the plain course, the river actively erodes its side by lateral erosion and swift frequently over the flat surface. The general tendency of running water is to flow in a straight course and if face any impediments on its path of flow it immediately tends to flow along an impediment free line. Within the limit of the flat surface over which it flows thus destroys the old meanders and consequently develops a new one. This is the initial process of the formation of **Ox-bow Lake**.

Flood plain: The areas inundated during peak flow period of a river is called **flood plain** zone of a river. During rainy season, the river water overtops the bank and submerged substantial neighbouring areas. The entire area consecutively so affected is filled with fresh alluvial soil and develops a broad flat plain. The flow velocity of the river within the confinement of the bank is relatively faster than outside of the bank. Most of the silt carried by the river is deposited along the entire area of its active flow. The areas along the bank receive fresh deposit every year is called **zone of new alluvium** and the areas receive deposits occasionally during high flood period is called **zone of old alluvium**. In the Ganga plain, the zone of newer alluvium is known as **Khadar** while the outer plain zone that receive deposits occasionally during high flood period is called **Bangar**.

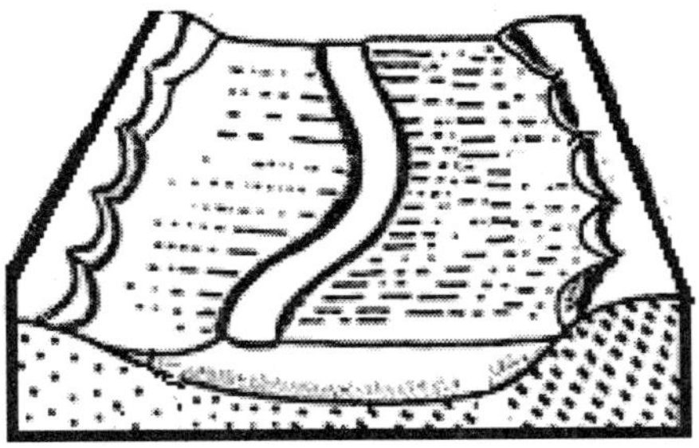

Fig. 60: Flood Plain.

Levee: Velocity of stream flow usually decreases backward direction. Sediments it carries along with the running water are successively deposited as it lost its carrying capacity due to the decrease of flow velocity and as a result, the river banks become progressively higher in comparison to adjacent areas. Usually river banks are submerged during peak flow period of a river. As the river water started receding it left silt, clay and sand at the river banks and through repetition of this process year after year, the river bank is raised above the level of the flood plain. The raised river bank is almost parallel to flow line of the river and portray embankment like longitudinal feature. This river bank deposit along the bank line is known as the **Levee or natural Levee**. Spectacular levee development is found along most of the tributaries of the Ganges and the Brahmaputra River.

Landform features developed in the lower course

The flow velocity of water decreases significantly towards the lower course of the river and therefore, it could not transport the sediments brought from the successive upper course of the river. In most of the rivers, the general tendency for deposition of sediments is quite conspicuous. Even mud and fine grained particles start getting deposited in its bed. As a result of such depositional activities of the river, the channel becomes shallow and wide. During peak flow periods most of the river banks are submerged under water and clay, silt and other finer particles easily overflow the banks. As the river water recedes, a large quantity of sediments based on their size and weight are deposited in large around the river banks. In consequence, a natural levee develops by the deposition of silt and sand alongside the channel and a flat plain subject to periodic inundation and deposition is built up. Since the river course is in graded condition, therefore, it facilitates depositional activities in the riverbed also.

Deltas

The flow velocity of a river ends where it meets the sea, a relatively stationary water body, with no velocity of water; the river deposited all the sediments irrespective of size and weight at the river mouth, the end of a river. The sediments are largely deposited at the river mouth along the seacoasts. As these are not transported into the deep sea areas, the sediments started accumulating at the mouth and gradually the rising river beds develop numerous distributaries to facilitate the discharge of river water. As a result of continuous deposition of sediments at the river mouth, the river naturally divides its course into some fresh channels to discharge water into the sea. The alluvial deposits around which a number of distributaries develop are triangular with the seaward bare. This triangular shape of the alluvial deposits in the river mouth resembles the Greek letter delta and accordingly the feature is named as **Delta**. Deltas are naturally a product of river deposition at the culminating point of a river and the sea. There are few preconditions for the formation of delta in a river mouth. These are:

1. The river must bring substantial amount of sediments along its flow.
2. The waves and tides of the sea should not be so powerful as to carry away all the sediments brought by the river. It means the sea should be comparatively calm and quite at the river mouth area.
3. The sea should be relatively shallow and contrarily deep sea is quite in congenial for the growth of delta in the river mouth.

The Ganges and the Brahmaputra together has built a massive delta near its mouth at the head of the Bay of Bengal. Delta is also found in the mouth of the rivers Mahanadi, Godavari and Krishna. On the contrary, delta formation is totally absent in the most of the west flowing rivers of India, Narmada and Tapi, the two west flowing rivers do not have any delta in its mouth. The tides and currents in the Arabian Sea are so stronger that no silt can be deposited near the mouth of the rivers. Moreover, these rivers flow through hilly areas and could not carry much sediment with them.

In general, the deltas are triangular but their actual shape varies from one to another in reality. On the basis of their varying shape, deltas are classified into four major types. These are:

(i) Arcuate type, (ii) Bird-foot delta, (iii) Estuarine delta, and (iv) cuspate or tooth-shaped delta.

Arcuate delta : This type of delta is more or less common and similar to Greek letter delta in shape. Its shape is like an arc of a circle

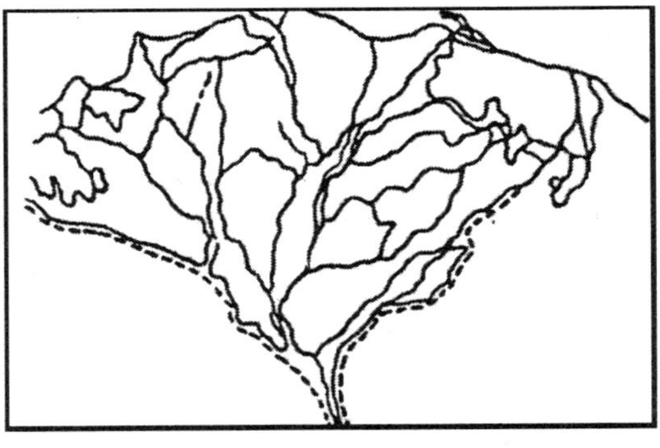

Source: Mitra, Kalpana, Sen, Jyotirmoy & Sengupta Purabi: *Physical Geography* (2014) Kalyani Publisher, Ludhiana, p. 320.

Fig. 61: Arcuate delta.

and exhibits a bow like shape. This type of deltas are fan shaped or actuate in form where the base of the delta positioned in the seaward side. Deltas of river Nile, Rhine, Hwang ho, Ganges etc. are arcuate in its shape.

(ii) Bird-foot delta: This type of delta is similar to fingers of human hand or foot of birds. This type of delta therefore, is called **bird-foot delta** or **digitate delta**. Mississipi River is formed due to the deposition of sediments primarily composed of limestone and fine-grained sedimentary rocks. The finer sediments enter into the sea up to a considerable distance as a result of slow velocity of the river. On account of the salinity of sea water most of the sediments are deposited along both side of the channel flow and finally give rise to finger shaped features. The distributaries thus deposited in both side of the flow results in the formation of **finger-shaped or bird-foot delta.**

Source: Mitra, Kalpana, Sen, Jyotirmoy & Sengupta Purabi: *Physical Geography* (2014) Kalyani Publisher, Ludhiana, p. 320.

Fig. 62: Bird-foot delta.

(iii) Estuarine delta: A long and narrow delta is formed when the river enters the sea only through a single estuary or mouth. As the estuary is filled up with sediments, a long normal delta is formed. As estuaries are open to the sea therefore, river sediments are often removed by the waves and currents of the sea. However, when the river mouth is submerged, a narrow linear delta is formed as a result of the deposition of the sediments. This type of delta is commonly found in the river mouth of Seine river of France.

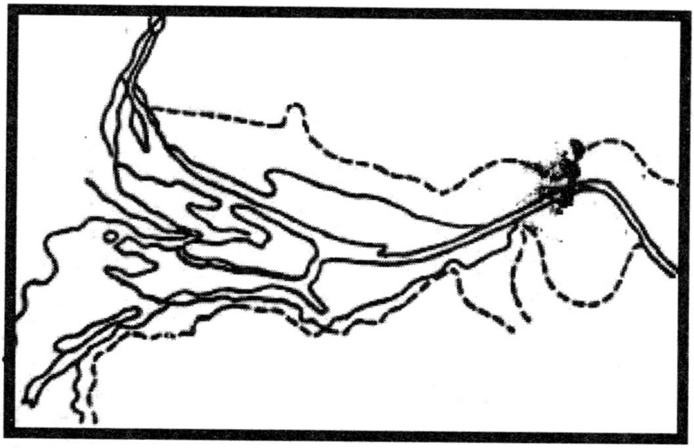

Source: Mitra, Kalpana, Sen, Jyotirmoy & Sengupta Purabi: *Physical Geography* (2014) Kalyani Publisher, Ludhiana, p. 321.

Fig. 63: Estuarine delta.

(iv) Cuspate or Tooth-shaped delta: When deposition of sediments takes place along the main channel and gradually project towards the sea. But the waves of the sea spread the deposits on both sides of the river mouth. Because of continuous attack of sea waves the deposition takes a curved shape on both side of the river and become concave towards the sea and become similar to a **tooth-shape**. This type of delta is found in the river mouth of Tiber River in Italy.

The landforms normally developed by river action can broadly be classified into two types—erosional landforms and depositional landforms. Gorge, canyon, rapids, waterfalls, potholes, river terraces, meanders; peneplains are the resultant landforms of erosional activity while alluvial fans, alluvial cones, natural levees, floodplains, ox-bow lakes, deltas etc. are product of depositional activities of a river. These features are conspicuous in different parts of a river namely upper course, middle course and lower course. Erosional landforms are usually found in the upper course of a river as the erosional activity is more intensive than the middle and lower course. Contrary to this, the lower course of a river possesses depositional landform following its extensive depositional activity in this part of a river.

Source: Mitra, Kalpana, Sen, Jyotirmoy & Sengupta Purabi: *Physical Geography* (2014) Kalyani Publisher, Ludhiana, p. 321.

Fig. 64: Cuspate or Tooth-shaped delta.

Characteristics of landforms associated with arid erosion

Wind is solely the most pervading agent of erosion in areas of arid climate. Wind as an agent of geomorphic change is not significant enough like water. Arid regions naturally lack rich vegetative cover, engulfed by extensive sandy soil surface and have cover of extensive base rocks. There conditions are congenial for wind action, which effectively performs erosion, transportation and depositional works in areas of arid climate. Wind action is also effective in areas having extensive sand cover.

Erosion by wind is performed primarily by three processes—**attrition, deflation and abrasion or corrosion**. Combined effects of all these three erosional processes give birth to various landform features in the arid regions.

Attrition: Extensive mechanical disintegration of rocks in consequence of high diurnal range of temperature and frost action. The wind as it carries the rock particles, they not only strike against the rocks standing in their way but also strike against one another. As a result of continuous friction they break down further and reduce to smaller size. Ultimately, these particles converted into sand and dust which covered extensive areas of desert. This process of gradual reduction in the size of rock particles by mutual friction is called **attrition**.

Deflation: In most of the desert regions sands and dust particles lost adhesiveness due to intense heat and dryness. Wind can easily lift these unconsolidated dust and sand particles and transport along its flow direction. The lifting movement is called **deflation**. Large-scale removal of sand and dust from a particular point of desert region may result either in the formation of depression called blowouts or in the exposure of the rocky surface lying beneath the layer of sand and dusts. In sandy deserts deflation hollows or blowouts of various sizes are formed and even wide extensive depressions are produced in the desert areas. Such depressions are found in the desert lying west of Nile valley, Kalahari Desert, West Australian deserts, Mongolian desert etc. In the sandy deserts deflation hollows of various sizes are formed and even basin like depressions may be produced under certain circumstances. Reportedly, the Quattara depression in western Egyptian Sahara has been excavated to a depth of 134 meter below sea level. These depressions favour the formation of Oases and in places saline lakes or swamps. It is conceived that if deflation begins at one point and continued for long it favours the formation of wind eddies and as a result, the hollow goes on increasing until the ground water table is reached. The final consequence is the formation of extensive depressions.

Abrasion or corrosion

Strong winds generally carry large quantities of sand and small angular rock fragments, which act as tools of erosion as they strike against rock surfaces of desert areas. Quartz is the important mineral of sand and is comparatively harder that most of the rock forming minerals. Such sand particles carried by the wind act as powerful sandpaper against the homogenous rock surfaces. Thus, the rocks are polished and smoothened as the sand particles and angular rock fragments strikes over them. In places where there are both hard and soft rocks, the soft rocks are entered faster and the surface become grooved and fluted. Conglomerate, and uncristallized rock composed of both hard and soft rock, as it comes under the influence of such strong wind, the relatively soft cementing materials are abraded at a faster rate leaving the relatively harder rocks less effected a left uneroded. It, in turn, takes the shape of a beehives and present honeycomb surface. The rock fragments, which are worn and polished by wind abrasion, exhibit one or more polished and fluted surfaces with sharp edges. This type of rocks in desert region is called **ventifacts or dreikanter**. Maximum abrasion, however, confined up to a little height of ground level as the wind connote lift sand more than a few meters height. Thus effect of abrasion reduces with the increasing height. As a result, the under-cutting of wind abrasion is more pervasive. In this way, the rock domes may be cut at the base of escarpments and the overexposed portion of the rock takes **mushroom** shaped blocks. This type of rock block is known as **gase** in Sahara desert. This type of rock blocks are also found in parts of western Rajasthan desert.

Source: Mitra, Kalpana, Sen, Jyotirmoy & Sengupta Purabi: *Physical Geography* (2014) Kalyani Publisher, Ludhiana, p. 364.

Fig. 65: Mushroom.

Some other landform features of arid region

Zeugen: Upstanding tabular rock almost similar to inkpot is formed because of wind erosion in arid regions. This type of landform features is called zeugen. It consists of an upstanding tabular rock undercut around its edges by wind abrasion. It is formed in areas where horizontally bedded strata occur with alternate composition of hard and soft layer. The hard rocks of the upper stratum are usually resistant to corrosion but as a result of the accumulation of dews in the joints which are freeze at night and remain protected from sunlight at day time and it comes under attach of mechanical weathering. The freezing process for a long period results in disintegration of rocks in and around the joints. Wind in turns remove disintegrated rocks from the joints and the process continues alike. Once the soft rock layer become exposed it becomes easier to remove rocks through erosion at a faster rate. In this way, the upstanding tabular rock masses remain upstanding over a throat-cut rock along the side and an inkpot like feature is developed. This type of feature in the desert region is called **Zeugen**. Zeugen may vary in height from a meter to about 30 meter.

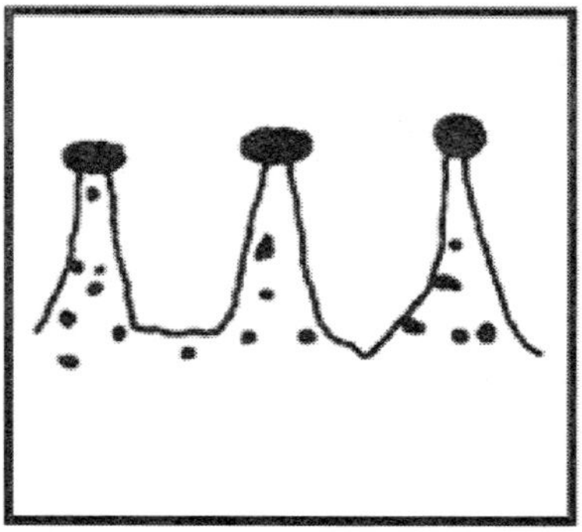

Source: Mitra, Kalpana, Sen, Jyotirmoy & Sengupta Purabi: *Physical Geography* (2014), Kalyani Publisher, Ludhiana, p. 364.

Fig. 70. Zeugen.

Yardangs: A distinctive type of landform develops in areas where layers of hard and soft rocks alternately stand in a vertical dimension on the ground. The vertical arrangement of the hard and soft rocks if are in a parallel direction of the prevailing wind, a typical feature comes into existence which is called **Yardang**. The development of Yardang is wholly attributed to wind abrasion. It consists of elongated grooves, which are cut by wind blown sand in the direction of prevailing winds. Removal of soft rocks open up a thorough way for wind to blow through the rock mass and aggravate more to enlarge the gap between the two vertical hard rock layers. These steep sided linear rock ridges separated by a corridor like depression develop in-group. This type of features are commonly found in Turkistan and part of central Asia. The gap between the vertical rock strata range from 2 to 10 meter while the height is usually less than 2 meter only.

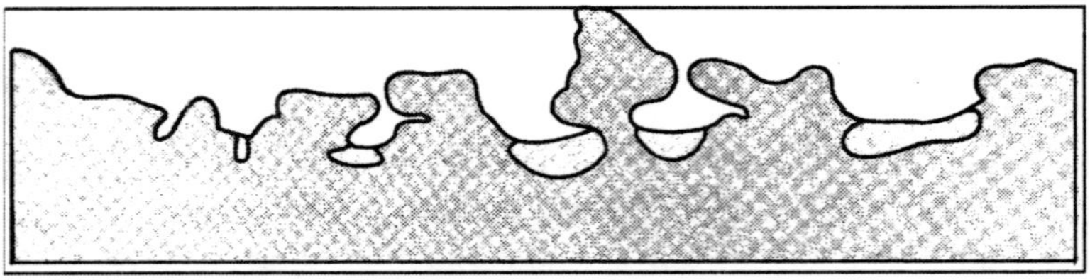

Fig. 67: Yardang.

Loess: Loess is a depositional feature developed in the desert areas, which is a product of windblown deposit, composed of fine-grained silt and dust. The loesses are developed away from the source region and sometimes outside of the desert. Loesses formed in the hot desert are called **desert loess or soft loess** and those are formed in cold region are called **cold loess**. Deposition of windblown sand and dusts take place either due to reduction in wind velocity or due to rain. The dust particles are so smooth and fine-grained that once these are deposited cannot be easily lifted by the wind. Loess is coherent but not cemented and hence is permeable. Extensive deposits of loess are found in many parts of the world. North European plain, Asiatic Steppes are congenial ground for the formation of loess.

Varying types of sand deposits

Sand particles, comparatively coarser and heavier and deposited by wind within the desert itself as the wind cannot carry these coarser sand particles to a long distance. On the basis of forms, the sand deposits can be classified as (a) Sand dunes, (b) sand drifts, (c) sand levees, (d) undulations, and (e) sand sheets. Sand dunes are more or less common among all these type of sand deposits.

Sand dunes: Sand dunes are extensive mobile heaps or pile of windblown sand. The formation of such sand piles is not dependent either on the form of the ground or any form of obstruction. These dunes are not static and free to move and divide subject to the prevailing condition of wind movement. They usually develop in different size and shape. Their precise form is dependent on the strength and direction of the prevailing wind movement. The shape of the dunes can be rounded or elongated and at time narrow sand pile. The dunes are found in varying size and under some conditions, the dunes may be of several hundred meters high and several kilometers long. Dune complexes are formed in places where groups of dunes develop the most essential condition for the formation of sand dunes are (i) abundant supply of sand, (ii) presence of unidirectional strong winds, and (iii) presence of some initial obstruction in the path of wind.

Sandy deserts are favourable ground for the formation of sand dunes and occasionally sand dunes may occur in the coastal areas having a sufficient supply of sands. The desert surface containing shale or limestone does not favour the formation of sand dunes; as such, grounds are not congenial to the formation of sufficient sands. Sands are found abundant in areas of surface formed of sandstones. Sand dunes are formed in varying sizes and shapes depending upon wind action, vegetation cover and supply of sands. On the basis of the size and slopes of sand dunes classified into two types (i) **Barchan** or **the crescent shaped transverse dune**, and (ii) **the seif** or **the longitudinal dunes**. The formation of transverse barchans is associated with large quantities of sand and relatively weak winds. There are seldom developed in the interior of great deserts rather they form along coastlines. The transverse barchan have their long axis almost at right angles to the wind direction. The slope of the barchan dunes is gentle towards windward side and steep towards leeward side. Sands carried by tend to fall from the crust of dune to the leeward side and is called **sand fall slope**. On account of the sand fall and the eddy action of wind, the leeward side develops a concave pattern of slope while the windward side is characterized by convex pattern of slope. Maximum deposition of sands are concentrated along the axis of the wind movement and gradually decreases outward margin from the axis as the wind velocity decreases outward from the prime axis of the wind movement. Barchans are therefore, crescent shaped. The average height of barchans is found to be 30-35 meter and their width ranged within 10-12 meters. Sizes of barchans depend on the supply of sand. The barchans tend to migrate downwind as the sand blows up the windward side and falls on the leeward side. But the migration of barchans in the desert takes place without distorting their shape. Sometimes, the progress of dunes are stabilized by rainfall and by the growth of grass and other plants.

Seif or longitudinal dunes: Longitudinal profiles are quite different from the barchans in both size and shape. These are developed as linear and long ridges parallel to the prevailing wind direction. These linear ridges extend for hundreds of kilometers in length and one a more kilometer wide with a height of several meters. Seifs are occurred in large parallel groups demarcated by narrow corridors between two adjacent seifs. These corridors usually remain sand free, which are called **Gassy** in Sahara desert. The Gassies swept by strong wind movement and thus keep it sand free. The strong wind movement blows out the sand from these passages and deposits the sand on either margin of the Gassies by eddy action.

The sief are well developed in the deserts of effective trade winds where the wind blows in a constant direction or reverse seasonally. Some geomorphologists opined that seifs are not solely the product of unidirectional wind movement. According to them, the longitudinal dunes of Sahara desert are developed as a result of cross wind movement and thus grew in length downward during periods when prevailing wind is parallel to the trend of seif chain. Barchans and seifs are rarely found together. In Thar desert of India although the barchans and seifs are found but they grow in different areas. Seif dunes are found in the western part of the Thar desert where the south-west monsoon winds are more powerful. Barchans are found in near Jodhpur and Bikaner area where they are extending for a length of 200 to 1000 meters and 180 to 600 meters in width.

Moreover, some other occasional features are developed along with the depositional process of sand in the desert region. Such features are sand shadow, sand drift, sand sheets, sand levees etc. Sand shadow is developed when wind movement face obstruction by the boulders, bushes, cliffs etc. in the path of the wind movement. The wind velocity decreases as it faces such obstruction and sands carried by the wind is deposited behind the shelter of the obstruction. Such a deposit is called sand shadow. A gap between two successive obstruction acts like a funnel through which sand is pushed forward and deposited on the leeward side of wind movement. Such a gap allowing sand to drift forward is called **sand drift**. A flat sand surface without having any topographic irregularity is called **sand sheet**. Such a sand sheet may have some ripples which resembles the sea waves. Selima sand sheet of Libyan deserts is nearly flat and covers an area of nearly 8,000 sq. km.

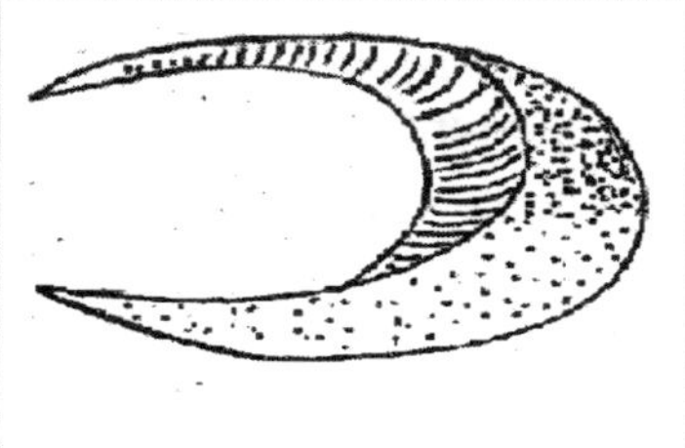

Fig. 68: Barchan.

Fig. 69: Seif.

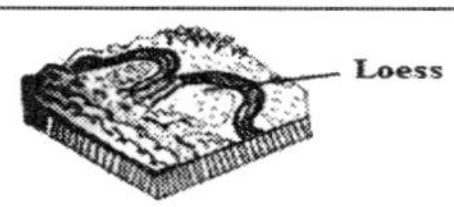

Fig. 70: Loess.

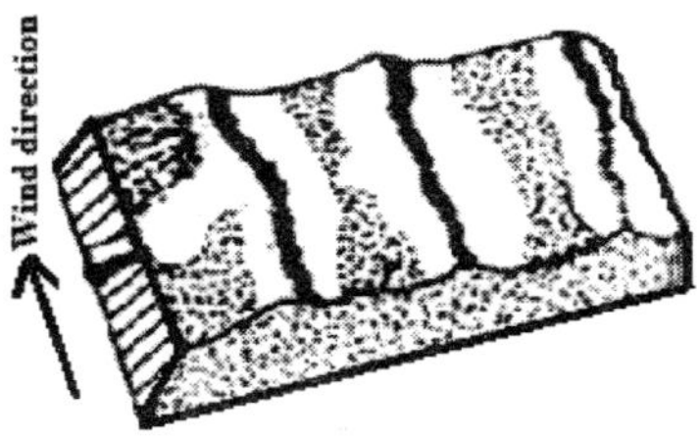

Fig. 71: Longitudinal Sand Dune.

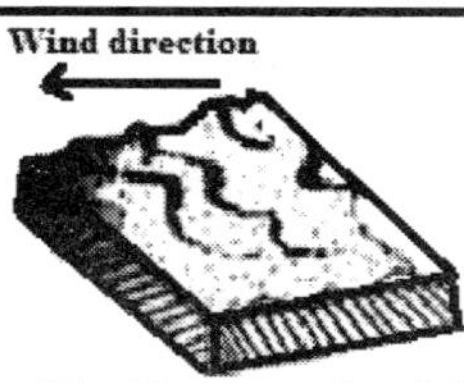

Fig. 72: Traverse Sand Dune.

Source: Mitra, Kalpana, Sen, Jyotirmoy & Sengupta Purabi: *Physical Geography* (2014), Kalyani Publisher, Ludhiana, pp. 371-372.

Sand levees are similar to a sand ridge with flat tops following almost parallel course of wind direction. This sand levee has no collapsing front like the seifs. These levees are formed from residue of dune chains. As the chain moves in downwind direction, the coarser and heavier sands are left behind as a distinct ridge. Such ridges are found in numerous deserts particularly tropical region. About 300 km. long and 1-3 km. wide sand levees of Egyptian desert region is one of the prominent example. These sand levees are also regarded as wholebacks based on its structural pattern.

Landform features of arid region resulting from water action

From climatic point of view all deserts are characteristically dry having a little rainfall in a period of several years. There may have no rainfall at all in a year, but occasionally there may be torrential rainfall particularly in the mountainous areas. In consequence, the streams, which are remaining dry for most of the time, experience sudden floods. Because of their spasmodic and impetuous flow Davis called them stream floods rather than stream. Stream floods occasionally give rise to sheet floods, i.e. broad a sheet of water suddenly flows downward irrespective to channel directions. Sheet floods result from heavy downpours on barren surface enriched with loose detritus. These are short-lived phenomena and undergone rapid drying up as a result of strong evaporation and loaded with dry detritus. These sheet floods flow for a short distances and most of the detritus carried are deposited in the downslope region in the form of alluvial fans in the foothill zone of mountainous region. Several fans like deposits of pebbles, cobbles and sands collide each other and develop a relatively flat surface, which is called **desert pavement**.

Some perennial rivers are also flowing through desert region having their origin in the rainy mountains located outside the desert. Nile, Colorado etc. are such few rivers flowing through desert areas with a consistent supply of water all throughout the year. **Bolson** is a typical feature usually developed in low-lying ground or basin surrounded by mountains. Most of the drainage are centripetal where Bolson developed. The mountainous river supply both water and detritus at the time of their flow and formed a temporary lake. With the rapid evaporation of water, the lake is transformed into a swamp and when completely dried up it becomes a level plain of fine sediments. Such a level alluvial surface is called **Playa** and if found in form of lake then such feature is called **Playa Lake**. Most of the playa lake remains water filled only after rainfall but most of them dried up within a short span of time after rainfall. The surface of the playa lakes are often covered with gypsum and salt precipitated from the water dried up after the downpour. Such salt covered basin is called **Alkali flats** or **Salinas** and these are found in most of the intermontane basins of North America and central Asia.

Pediments

A sharp break in slope usually developed at the meeting point of mountain and the adjacent plain of arid and semi-arid mountainous region. The plains are necessarily gently sloping which the mountains have steep slope. The area of gentle slope lying between the mountain and the playa is called piedmont. The upper portion lying in attachment of the mountain is free from having any sedimentation and exposed as an eroded bedrock surface. It is formed as a result of extensive degradation in the mountainous slope zone. This part of the surface exposing the bedrock is called **pediment**. Pediment is the upper part of the piedmont lying at the close attachment of the mountain. On the contrary, the lower part of the piedmont which possesses depositional detritus in large over the surface is known as **Bajada**. Bajada is characteristically an aggradational feature and contains large quantities of debris derived from erosion in the upslope areas. Occasionally, pediment may have a thin cover of debris primarily composed of pebbles and granular detritus. The slope over bajada and pediment is normally gentle. But there may be a sharp break in slope at the meeting point of pediment and mountainous front.

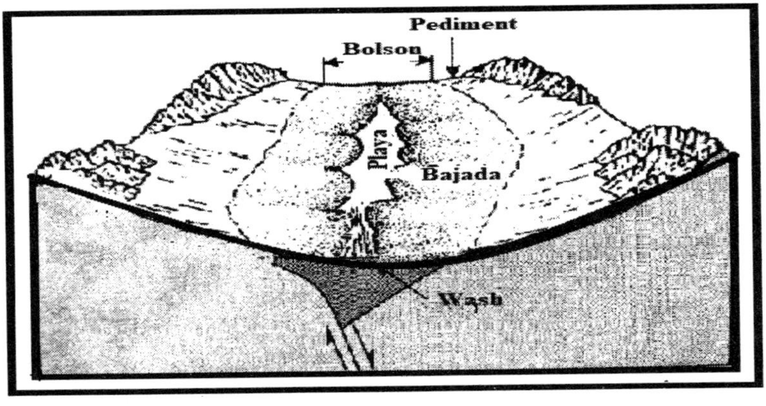

Fig. 73: Typical assemblage of mountainous desert forms (after A.L. Bloom, 1992).

The term pediment is meant to designate the rock-floored plains adjacent to or interpenetrating mountain masses in desert regions. However, these pediments are developed by planation of the mountains at altitudes dependent upon those of the interior basins into which the mountain drain. The term pediplain is generally used to describe a series of coalescing pediments. On the other hand, bajada is formed by coalescing alluvial fans. At the time of rain, the mountainous streams bring down large quantities of sediments composed of pebbles, boulders, sand, clay etc. and deposit them near the foothills to give birth of the alluvial fans. The bigger stream develops broader alluvial fans of low gradient while the smaller streams form smaller alluvial fans of steep slope. Most of such streams are ephenural and hence frequently shifted their course over the fans making the bajada almost undulatory. Several theories regarding the origin of pediment are put forwarded by a number of geomorphologists. The prominent among them are Blackweldev, Johnson, Davis, Bryan, King etc.

Pediplains and Inselberge

An inselberg is a steep sided, isolated hill composed of extremely hard and resistant rock rising like an island over the flat plain of desert region. The island like residual hills found in the deserts is called **inselberge** in German, the literally meaning of which is island mountain. Later on, the term becomes a common term for all forms of residual knobs on pediments in arid and semi-arid regions irrespective of their origin and location. The inselbergs are steep sided residual hills of hard resistant rocks locating amidst extensive flat surface of the pedipalin. Rocks of surrounding areas have been cut and removed by large scale weathering, erosion and deflation by wind action. The inselbergs are similar to the **monadonocks** formed in the peneplain. The actual shape of inselbergs is determined by the nature of rocks. In most places, the inselbergs are composed of granites and gneisses.

There is a considerable difference of opinion amongst the scholars regarding the processes of the development of inselbergs. Earlier, it was assumed that inselbergs are essentially a product of wind erosion only. Desert plains with the presence of residual inselbergs were considered the last stage of arid erosion cycle. However, later studies confirmed that wind alone is not capable in reducing such extensive areas to low-level plains by corrosion and deflation. Besides wind, water erosion must play a vital role in reducing such extensive areas into a flat plain. Not only in desert alone, is the inselberg landscape extensively found in the Savannah areas of Africa. The presence of inselbergs in the semi-arid Savannah areas are considered as one solid proof of the fact that inselbergs were formed during Pleistocene ice age when there prevailed

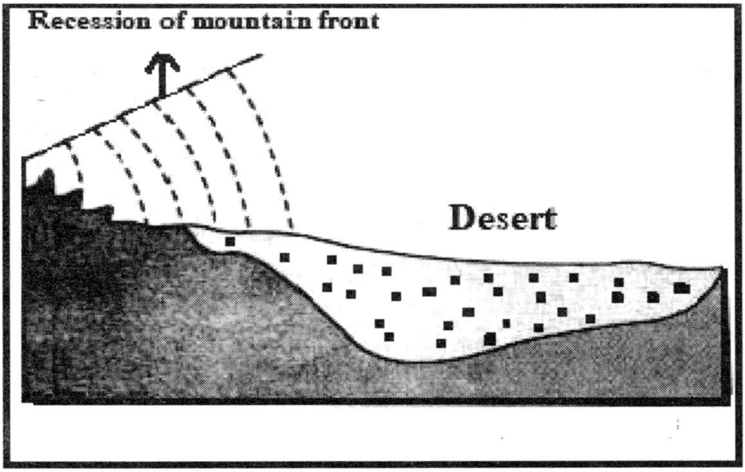

Fig. 74: Formation of Pediment (after Lawson).

humid climatic condition and favoured river erosion during that period. Peel (1966) claimed that many of desert landforms appear to be relics of erosional activity during that of the humid period. Cotton (1942) opined that the inselberg of the Savannah may have been produced largely by lateral erosion of rivers and the inselbergs of arid and semi-arid areas of Africa may have pedimented landscape in which residual hills similar to the inselbergs rise above the coalescing pediments. According to King (1948), the disintegration of rocks by weathering to such a depth of 1000-1500 ft. is quite impossible. Accordingly, the first requirement is the presence of granite or gneisses, which are seldom effected by chemical weathering. The second important requirement is the presence of rectangular joints in the rocks, which determines the drainage pattern.

Inselbergs are formed in the second cycle of erosion of a rejuvenated landscape. At the end of the first cycle, the drainage pattern make adjustment with the rectangular joints of the rocks and as rejuvenation starts, the rivers accordingly started cutting downwards along the joints. As a result of this process, the interfluves assume the slope of steep sided inselbergs. Many more concepts and hypotheses are forwarded by numbers of scholars regarding the origin of inselbergs. However, controversies still exist particularly on the aspect of height of inselbergs, which are found to be 1000-1500 feet height.

Fig. 75: Inselbergs.

Characteristics of landforms associated with Karst region

Limestone, dolomite, rock salt, Gypsum beneath the land surface and there way produce some solutions, which are not found in a well-exposed condition. Almost all the states in the United States although possess such soluble rocks but largely fails to develop well exposed Karst topography except some converses which do not constitute a well developed Karst topography. As such coals bad in south-western New Mexico possesses a members of caverns

The word Karst is a term usually applied to limestone areas, which possesses a specific topography different from others. Karst topography usually formed from dissolution of soluble rocks like limestone, dolomite and gypsum. Underground drainage system is liable for the formation of few characteristic features in the Karst region namely sinkholes, doline, cover etc. The development of all Karst landforms requires the presence of rock, which is usually dissolved by surface water or underground water. Understanding the Karst topography is important because ten per cent of the Earth's surface is occurred by Karst landscape. Karst region occurs in temperate, tropical and in polar environment. The significance of rock solution is attached with the reduction of landmasses.

At least four conditions are essential for a distinct and maximum development of Karst topography. The first one is the presence of soluble. Rock at or near the surface, preferably limestone. Dolomite and chalk is also soluble but other rocks usually lacks one of the other prerequisites. Secondly, the layer of such soluble rock should be dense, highly jointed and thinly bedded. When rock is highly process and permeable throughout, then rainwater will be absorbed enoromous and as a result move throughout the whole rock mass rather than be concentrated along specific lines of movement. For this reason, the Karst topographic features are not well developed in the chalk region of England and France. The third condition is the existence of entrenched valleys below uplands underlain by well-jointed and soluble rocks. This type of structure favours the downward movement of groundwater through the rocks. Solution in limestones acts as conduits for the movement of groundwater through the rock strata. Effective water circulation is a vital prerequisite in order to supply soluted water through the rock strata. Fourthly, the region must have at least moderate rainfall. It is significant that nearly all the important Karst regions are located in areas of having moderate to heavy rainfall. Contrary to this, arid and semi-arid regions do not exhibit distinct development of Karst topography. Existences of Karst features in such areas are relic features formed during more humid climate conditions.

Significant Karst Areas: Important Karst regions are normally found in areas where limestones underlie the surface. Moreover, in some areas, the rocks are dolomitic limestone and under some favourable conditions, the Karst topography develops in a very distinctive manner. Eventually solutional features may develop in regions predominated by soluble rocks like gypsum and rocksoil. However, limestones over the earth's surface are abundant and Karst topography is found well developed in these regions. Well-developed Karst topography is found in many parts of the world. The important regions are great valley region of Pennsylvania, Maryland, Virginia and Tennessee of U.S.A, cause region of southern France, Spanish Andalusia, Northern Yucatan and Tabasco of Mexico, northern Puerto Rico, western Cuba, central New Guinea, South-east Asia, New South Wales and Western Australia. The development topographic features however depend on varying conditions, such as lithology, structure, light of water table and climatic conditions. This is well illustrated by some significant difference between tropical Karst and the Karst of middle latitude zone.

Insignificant Karst Areas: There are innumerable areas over the earth's surface where Karst features are present but in noway dominate the whole landscape. This may be attributed to the conditions such as topographic youthfulness and other essential conditions prerequisite to ideal development of Karst topography. However, such areas are dominated by soluble rocks such as limestone, dolomite, rock salt, gypsam beneath the land surface and there way produce some solution, which are not found in a well-explored condition. Almost all the states in the United States although possess such soluble rocks but largely fails to develop well exposed Karst topography except some converses, which do not constitute a well-developed Karst topography. As such, Carls bad in south-eastern New Mexico possesses a numbers of caverns, a typical Karst feature beneath the plateau area but largely fails to develop other associated Karst's features. The chalk region of England and France is another specific example of an area which exhibits some Karst features, but the assemblage is too incomplete which in no way can be categorized as major Karst area. The chalk normally does not have well-developed joint systems hence downward movement of water is restricted and therefore localized solution and concerned features are not conspicuous in these areas.

Topographic features of Karst region

It is quite true that all the characteristic features discussed here are not found everywhere in Karst region. An attempt is made here to describe the important features that may be found at one place or another. Some of the Karst features are not topographic in character but they are closely associated with the topographic form and certainly need discussion to unearth the facts of Karst topography.

Terra Rossa: Rainwater mixed with atmospheric carbon dioxide and organic CO_2 which is generated in the ground by decomposition of plant debris, because active relevant agent and dissolves and disintegrate most of the carbonate rocks at or below the surface. The amount of dissolution of carbonate rocks depends on few factors like (1) Temperature, (2) magnitude of atmospheric content of carbon dioxide, organic carbon dioxide and chemical composition of carbonate rocks (calcium carbonate, magnesium carbonate), (3) Joints and fracture of the rocks, (4) rapture of flow of ground water, and (5) time-span of conduct between solvent and carbonic rocks etc. Temperature and dissolution of rocks always exhibit a positive relationship means higher is the temperature, higher is the dissolution and lower is the

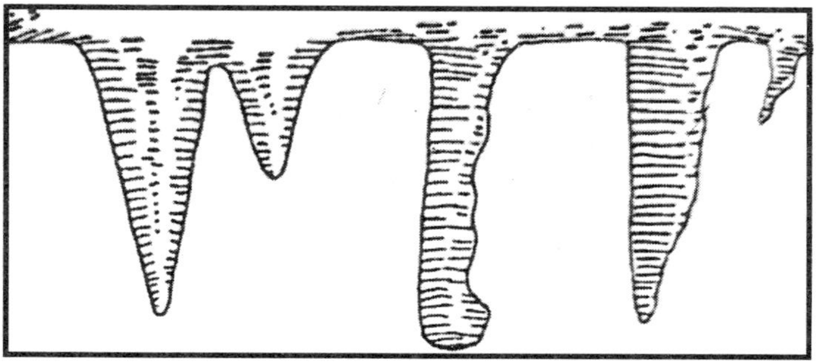

Source: Mitra, Kalpana, Sen, Jyotirmoy & Sengupta Purabi: *Physical Geography* (2014) Kalyani Publisher, Ludhiana, p. 418.

Fig. 76: Terra Rossa.

temperature, lower is the dissolution. The temperature condition acts, as one dominant factor is case of dissolution process of carbonate rocks. The descending ground water usually leaves a residue of red and clayey soil enveloping the surface and extending down into opened joints. This type of material is usually found on moderate to gentle slope. Contrary to that, material quickly comes down along the steep slope and remains totally absent in areas of steep slopes. This weathering residue as left at the surface are called terra rosa, which means red residual, soils or earth.

Lapies: Lapies are developed in limestone areas where relief is considerable. Such surfaces are bare of terra rosa and exposed on etched, pitted, ground and rugged in character. Lapies are characteristically fretted and fluted topography narrated by small rills and gullies along with minor ridges and deep cliefs. Lapies is a French Term and is known by different names in different parts of world as. Clitns or Rykes in North England, Karen in Germany, and Bogaz in Yugoslavia etc. A variety of lapies in which the solution grooms are rather small and closely speed is called rill lapies.

Conducting massive study on Karst region Cvijic (1924) has described the amazing diversity of surface and form, which are found well exposed in the Karst region. Accordingly, lapies are formed chiefly on the outcrops of naked rocks. The formation of lapies is dependent on some essential factors like rock composition, texture, structure, surface scope and amount of vegetative cover. These are rarely developed on horizontal rocks where dominance of sinkholes is prevalent. It is also

Fig. 77: Lapies.

agreed by some others that lapies develops under a soil and vegetation cover and is exposed later as a result of acceleration of corrosion and deforestation. In some Karst areas where terra rosa has been removed there exposed the true lapies. Lapies are seen well exposed in comparison to other Karst's features. Minute differences in rock solubility, permeability, jointing, bedding and other physical attributes influence the rate and localization of solution by descending water. Formation of lapies primarily depends on all these attributes. Lapies are formed due to corrosion of limestones along the joints subject to the exposure of limestone at the ground surface, as carbon dioxide content, rainwater proceeds over a region of lime stone rocks, it begins to dissolve.

Certain portion of the rock: The joints are successively widened by solution and the surface become fretted and fluted. Most of the lapies are formed in this way.

Sinkholes and associated features: Sinkhole is the most common and widespread topographic features in a Karst terrain. Rainwater containing atmospheric carbon dioxide dissolves with limestones and other carbonate rocks along their joints. In turn, develop various types of sinkholes. Topographically, a sinkhole is a depression that varies in depth from a mere indention of a few centimeters to 10 meters. The average depth however confined within 3 to 10 meters only. In along, they range from a few square meters to a few acres. The most general form is funnel-shaped and upward open part is comparatively broad along with some variations in size, shape and depth. Fundamentally, the sinkholes can be grouped into two types, those that are developed slowly down and by solution beneath a soil mantle without producing any physical disturbance of the rock and those that are produced by collapse of rock above an underground void respectively. They are referred as solution sinks and collapse sinks. Both are products of solution. Sinkholes are usually of two types (i) funnel-shaped sinkholes, and (ii) Cylindrical sinkholes. Gradual enlargement of sinkholes because of continuous dissolution of limestones results in the consequences of attached sinkholes. Such sinkholes are known as 'Swallow' hole. In places some swallow holes are further enlarged under effect of continuous solution and turn into a large depositions, which are called 'dolines' in the Karst region. Collapse of some portion of upper surface of solution holes due to the formation of stalacities below the ground surface, develops a feature 'collapse sinks'. Areal coverage of dolines usually ranges from few meters to 1000 meters and the depth ranges from a few meters to 300 meters in average.

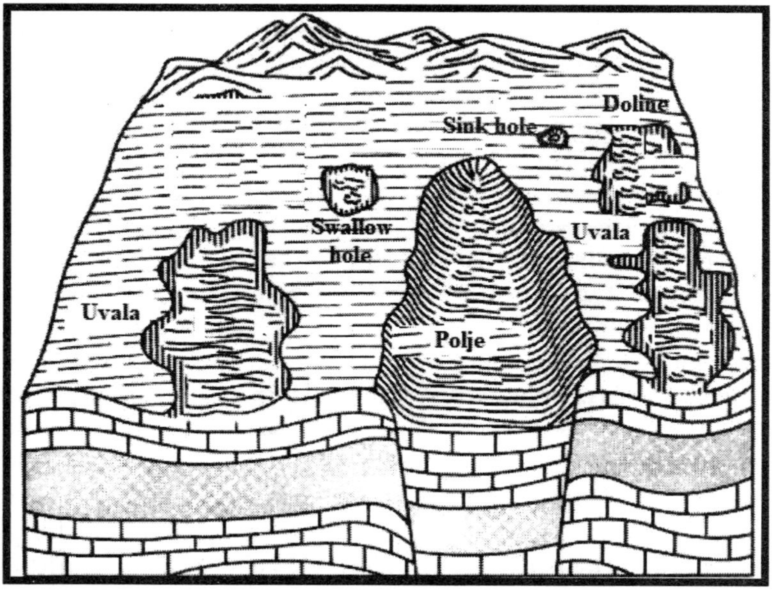

Source: Singh, Savindra: *Physical Geography* (2010), Prayag Pustak Bhawan, Allahabad, p. 276.

Fig. 78: Swallow hole, Sink hole, Uvala, Doline and Polje.

An almost similar feature to doline is the 'solution pan.' It differs from doline in areal coverage and depth. The solution pan is much shallower and may embrace a much layer area. The solution of Indiana county of U.S.A occupies an area of more than 30 acres. Eventually, the floor of some dolines sometimes plunged due to deposition of clay, which in turn prevent the downward water percolation and turn the doline into a lake like feature, which is called Karst lake or sinkhole ponds. Another feature termed as 'uvula' is formed under numerous factors. These are : (i) due to presence of several dolines as a result of continuous solution and enlargement of dolines, (ii) Due to collapse of the upper roof of large cavities developed underground, (iii) due to the merging of several sinkholes. The structure of Uvala merely depends upon the underground cavities and rock strata. A special type of feature is found along with some collapse sinkhole, which is termed as Karst window. The tern 'Karst window' has been applied to an unroofed portion of an underground stream course. Karst windows are formed due to collapse of a portion of upper surface of a sinkhole or any sub-surface feature. It provides a chance or an opportunity to observe the features developed in sub-surface zone. Karst window may vary from a mere peephole to one of considerable size. Some Karst windows have become greatly enlarged and possess alleviated floors across which the cavern stream flows as like surface streams.

A somewhat typical form of sinkhole is the Polje. An extensive, larger than dolines in size and shape is called poljes. Poljes are depressions of various forms and origin with a flat floor and steep enclosing walls, which owes its existence to solutional modification of downfaulted or downfolded limestone blocks. A polje may be similar to uvala to some extent but differs from it in origin and extent. Poljes are characteristically closed basins with an elliptical shape covering an area up to 258 square kilometer. This kind of topographic feature of karst region is normally found in Yugoslavia and Jamaica. Livno polje of western Baskan is the largest one covering an area of nearly 512 square kilometer. The poljes are believed to be formed as a result of downfolding and downfaulting of limestone areas due to earth movements. The landforms thus developed are then modified by solution work of water.

Other associated features of karst plain

A horizontal or gently dipping limestone strata having a sizeable number of sinkholes and other associated features is called a karst plain. Almost all the valleys possess sinkholes and swallow holes in any form accompanied by other features common to karst areas. In such a plain, the surface drainage systems develop various types of valleys and typical landform. The sinkholes are very effective in collecting surface run-off. Larger streams usually flow over the sinkholes sharing water with the subsurface stream linked with sinkholes, but small streams lose their waters to underground routes through the sinkholes. This type of feature is called sinking cracks. The point through which water goes downward is called sink. The water of small rivers disappears through a single sink while water of larger streams disappears through many sinks. Sinking creeks constitute an important aspect of karst terrains in relation both to their surface form and to the development of caverns beneath a sinkhole plain. Lost in southern Indian flows some eight miles underground before it reappears over the surface.

Rarely used or mostly unused valley below the surface where the streams sink is called its dry bed. Most of the beds become active during periods of heavy rainfall when shallow holes and other connecting underground links are unable to accommodate all Re-storm waters. A surface stream, which eventually disappears in the limestone region through a shallow hole or sinkhole, is called a blind valley. The surface continuation of the dry bed below the sink through long continued disuse becomes obscured. The blind valley may become a temporary lake when the swallow hole become unable to accommodate the increased volume of water accumulated as a result of heavy storm. Another type of blind valley, spectacularly distinct from that which usually reward a special name, is the solution valley or karst valleys. It normally develops upon the process by which a limestone terrain may expand at the expense of a region that originally was drained entirely by surface change and contains topographic characteristics developed by surface drainage. Surface drainage or the streams develop U-shaped valleys in their course on limestone formation. Such extensive U-shaped valleys developed on limestone are called solution valley or karst valley. The life and existence of karst valleys are uncertain as it becomes dry as a result of disappearance of water flow through the shallow or sinkholes. Sloans valley, Logsdon valley, Elk spring valley, Horse cave etc. are good examples of karst valley.

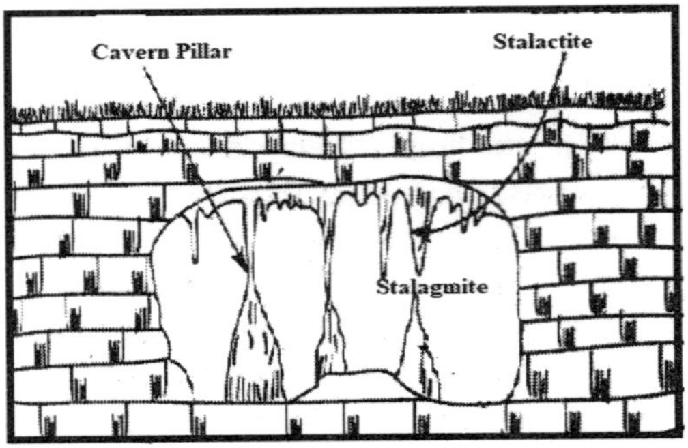

Fig. 79: Cavern Pillar, Stalactite & Stalagmite.

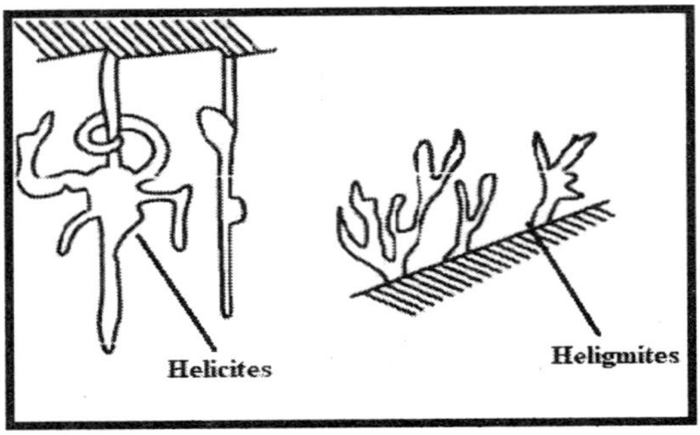

Source: Khullar, D.R.: *Physical Geography* (2014), Kalyani Publishers, Ludhiana, p. 416.

Fig. 80: Helicites and Heligmites.

Caverns and associated features: Generally, a cavern or cave can define as a natural subterranean runway. It may have vertical and horizontal extension and it may be found in one or more levels. It may extend both vertically or horizontally and may occupy one or more levels. Caves or caverns are the most significant landforms produced by erosional work namely corrosion, solution and abrasion by groundwater in limestone regions. Pure thickly bedded and massive limestone lithology is essential for the formation of larger caves. While smaller caverns are confined only along the joints and bedding planes.

Cavern possesses a variety of depositional features. The most striking features are developed by accumulation of calcium carbonate on the ceilings, walls and floors of the cavern. Travertime is a term generally applied to the many forms. Davis suggested that dripstone be used to include forms developed by water dripping from the cavern roof. This would include the downward extending stalactites, the upward growing from the floor stalagmites and the columns and pillars produced when the proceeding forms grow together. The water charged with the solution of limestone falls along the roof of a cavern drop by drop. During the time a drop is hanging from the roof of the cavern, some of the water evaporates leaving behind a deposition of lime. This deposition of lime grows downwards from the roof of the cavern like a hanging pillar. This kind of feature is known as stalactite. This hanging pillar is broad at the roof and become smaller or pointed at the downward extremity.

A stalagmite, on the other hand, is similar to stalactite pointing upwards from the floor of the cave. The drops of water that fall on the floor of the cavern after percolating down the roof also contains some lime which is left on the floor after evaporation. The lime thus accumulate on the floor begins to rise upward from the floor like a pillar. This is known as stalagmite. On the other hand, a term helicite is more often used to indicate a feature with verity of shape and size. It is unusual in that its growth does not necessarily extend along vertical limes. Its individual parts may grow upward, horizontally, obliquely or in curves. During the formation of helicites, it apparently defy the effect of gravity. Regarding this, the most logical explanation is forwarded as they develop where water is not entering the cave in sufficient quantity does give rise to drops that fall. In presence of sufficient water flow the surface remains wet then growth of helicite will not be possible. Growth of the individual parts of the helicite depends upon the crystal axes of calcium carbonate and may develop in any direction.

Natural tunnels and bridges: Natural tunnel is a subterranean cut-off or subterranean stream piracy. Natural tunnel is located in Scott county of Virginia. This natural tunnel is unique because it serves as a railway tunnel. It is 900 feet long, 130 feet wide and 75 feet in height. It is believed that this tunnel developed as a result of subterranean stream piracy, resulting in the diversion of surface waters from a higher topographic level on one side of purchase ridge to a lower topographic level on the other side of the ridge. Another explanation of natural tunnel is that is proved in consequence of the development of a subterranean cut-off by stock creek through the ridge and thus developed a natural tunnel beneath the ridge. Collapse of the roof of a natural tunnel will in time reduced the length, which in turn developed a natural bridge. Since natural bridges are developed in many ways, those formed as a result of solution by groundwater is called karst bridges. The most spectacular karst bridge is located in Rockbridge county of Virginia. Woodward (1936) forwarded a theory on the formation of natural bridge. According to this theory, this natural bridge was produced by subterranean stream piracy. According to the study conducted on the James River, the upper part of present cedar creek was once a part of the drainage of eastward flowing plague Run, The lower part of the valley of cedar creek was occupied by a short stream, which was earlier a part of present cascade creek. Both streams were tributaries of the river. There may have been a difference in elevation measuring about 70 meters in the two tributaries of the James River. The headwaters of plague Run were diverted through the underlying soluthe limestone to cascade creek. Thus an underground tunnel was formed which was in time de-rooted except for the parts remain as natural.

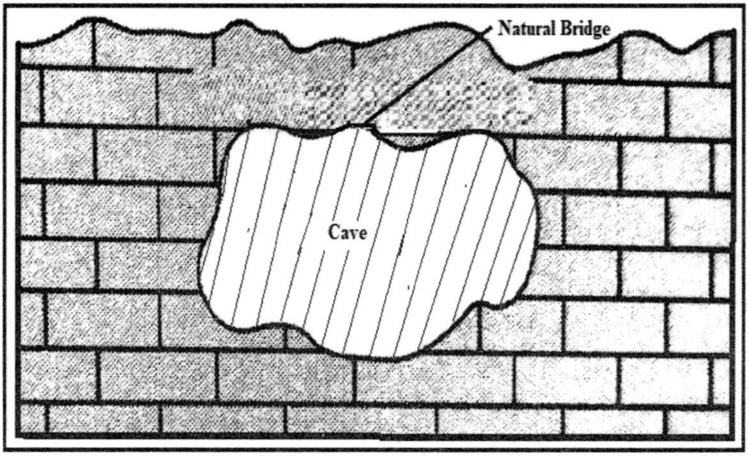

Source: Khullar, D.R.: *Physical Geography* (2014), Kalyani Publishers, Ludhiana, p. 415.

Fig. 81: Natural Bridge and Cave.

Glacial Landforms

Glaciers are the solid moving mass of ice. Like rivers, the glaciers perform three types of work—erosion, transportation and deposition. Glaciers are, in fact, most powerful agent of erosion. Glacial erosion takes place from the movement of ice sheet over the rock floor. In its movement, the rock fragments embedded with ice at the ice rock interface works as a basic tools of erosion. Around with sharp angular rock fragments of varying sizes the glacier as it moves under pressure polished, scratch and cut grooves in the rock surface. This process of glacial action is called **abrasion**. The nature of abrasion differs on the basis of its erosional activity and the debris it produces. Rock flour or sandy debris usually develops a polished surface while gravel a boulder size debris produces striation on the rock surface and sometimes it develops grooves on the surface. Larger size of striation is named as **grooves**.

Plucking or joint block removal a quarrying is another mechanism of glacial erosion. In terms of effectiveness, plucking is more important and erosive than abrasion. Block removal takes place in the well jointed rocks, which have already been loosened by chemical and frost action. An un-broken rock is resistant and relatively harder than the ice block. Therefore, the rocks need to have some shattering, cracks or fracture to undergo block removal by glacial activity. In most cases, existence of fractures either tectonic or developed by pre-glacial frost action is essential. There is also evidence that glacial ice can reinforce these fractures by the process of dilation or by surface drag. In the temperate region, particularly, the freeze thaw process also oscillates the development of cracks and fracture in the rock blocks. Plucking or block removal usually is more effective in areas having jointed rock blocks (Quartz, mica and schist) than the rocks which are not well jointed (Unjointed state).

Glaciers can transport large quantities of debris over considerable distances. Demorest (1942) considered that there are four types of glacial flow which he designated as gravity flow, obstructed gravity flow, extrusion flow and obstructed extrusion flow. Gravity flow takes place in areas where the gradient of the valley floor below the ice stream is sufficient for glacier to overcome the frictional retardation on the floor but does not allow producing an ice cascade or falls. Under these conditions, an ice stream can flow with a line of maximum velocity. Wherever, the flow of ice is regarded by some obstacle such as stagnant ice, or topographic obstruction. It tended to override the obstruction by way of shearing. Demorest called this process **obstructed gravity flow**. Extrusion flow takes places where the bedrock is basined and thickness of ice is extensive. Because of differential pressure within the ice mass, the basal portion of the ice started flowing down. Thus velocity of ice is more in the basal portion of the ice started flowing down. Thus velocity of ice is more in the basal part than near its top. Demorest designates such type of ice flow as extrusion flow. Any form of obstruction in the path of true gravity flow is called **obstructed extrusion flow**. Demorest believed that extrusion flow is more important in the movement of ice cap where motion is more pressure controlled than topographic control. The slopes or gradients on the ice surface determine the direction and velocity of an ice cap is not much dependent on the configuration of the topography upon which it rests. The topography beneath the ice cap may help to localize movement and also determine the velocity of movement to cause intensive erosion.

Cirques and Associated features

The term cirque was first used by French scholar Jean de Charpentias in his study of the Pyrenees in 1823. Cirques are the semi-circular steep sided depressions found at the head of valley glaciers. Snow accumulates in the depressions and condemned to form cirque glacier, which flows down slope to feed the valley glaciers. Cirques developed on amphitheatre shaped rocks similar to deep-seated arm chair shape. These depressions are filled with heavy glaciers. In the absence of glacier ice, cirques basins are more often occupied by small lakes called **tarns**. An empty cirque usually displays three different parts—a headwall, basin and threshold. A cirque headwall is steep and talus free at its base and the height may range within 2000 to 3000 feet.

The floor of a cirque is basin shaped which extends forward from the headwall and terminates at the bedrock riser called **threshold**. Cirque basins are in general like a small lake which is called **Cirque Lake** or **tarns**, a term used in the British Isles. Cirques vary in plan from simple, sub-circular outlines to compound ones, which are similar to many scallops. Compound cirques represent a more advanced stage of development and usually formed as a result of headward extension at varying rates in different parts of a cirque headwall. Differences in cirque characteristics is indicative to their stage of advancement in a cycle of mountain glaciations (Hobbs, 1910a, 1910b). Cirques are more common in some regions and distinct than some other areas, even under favourable climatic conditions and topographic congeniality. The necessary favourable conditions for cirque development are (i) rather wide spacing of pre-glacial valleys so as to favour expansion without intersection of adjacent cirques at an early stage, (ii) sufficient snowfall to form extensive snowfields and glaciers but not sufficient enough to form ice-caps, and (iii) fairly homogenous rocks which permit cirque extension equally in any direction (Thornbury, 1954).

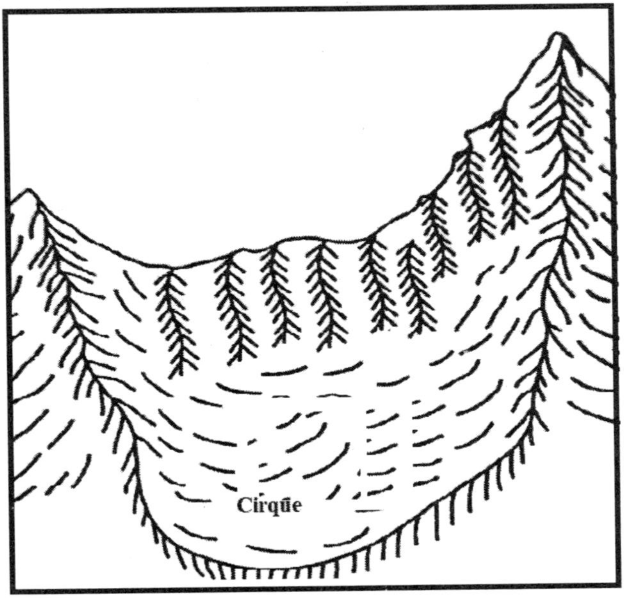

Source: Khullar, D.R.: *Physical Geography* (2014), Kalyani Publishers, Ludhiana, p. 458.

Fig. 82: Cirque.

Glacial Trough: Glacial trough is the most distinctive topographic feature in a glaciated mountain. The head of a glacial trough is attached with the lower edge of a cirque threshold. Most glacial troughs were originally stream cut valleys but as glaciers altered them greatly, therefore, the cross profile of glacial trough or the long profile does not confine within stream carved valleys. Trough characteristics are commonly found in areas only of extensive glaciers. There is usually a conspicuous drop of glacial trough from a cirque threshold to the floor of a glacial trough, which is called a **trough headwall**. Trough headwalls are conspicuous where more than one cirque contributes ice to a glacial trough. The trough floors descended over a series of glacial steps and these steps are usually more common in the upper part than in the lower part. Each step has three component parts—a riser, a riegel and a tread. Down valley end of each step and the ascent to it from below is called a **riser**. A sort of rock bar at the top of and just back of a riser is called a **riegel** while the relatively flat floor of a step is known as **tread**. A reverse up valley slope resulting in the formation of a basin give birth to a lake like feature. Following similarity of arrangement to beads on a rosary, such features is named as **Paternoster Lake**.

It is more often said that thickness of glacier will increase where the trough is narrow and it will upsurge the glacial abrasion. It results in greater trough deepening. Steps in a trough are attributed to the effects of varying rock hardness where rivers and associated riegels mark hard rock bars across a glacial trough. It has been suggested that steps represent pre-glacial irregularities or nick-points. The risers are not stationary. They migrate up a trough and at times disappear as the glacier persisted in it far long duration. Well-jointed or fractured rocks usually favours glacial plucking and eroded the massive rocks. Glacial valley differs significantly from un-glaciated valleys in mountainous areas. The glaciated valleys are U-shaped in form while the un-glaciated valleys are usually of V- shaped. The shape of the valleys are determined by the thickness of glacier, varying lithology and structure of the bed-rocks over which the trough is cut.

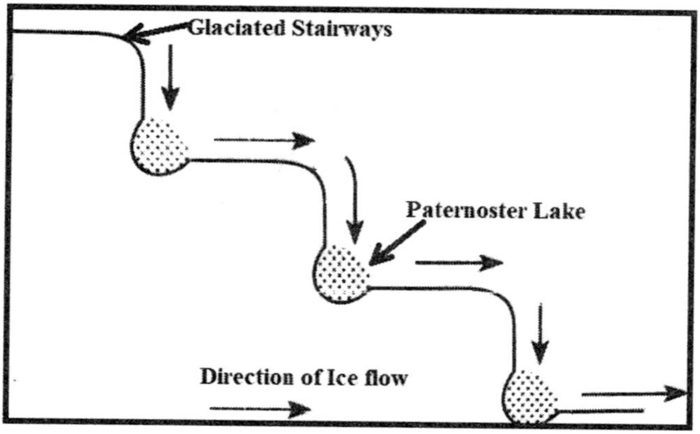

Source: Singh, Savindra: *Physical Geography* (2010), Prayag Pustak Bhawan, Allahabad, p. 276.

Fig. 83: Paternoster.

Hanging Valley: The valleys are straighter by the process of abrasion of spurs projecting from the sides of the original stream. The truncated spurs normally are found in different degrees of truncation. Some of the truncated spurs are practically cut and some others may be completely truncated. Along with the truncated spurs another feature is developed which occur along the valley sides. The tributary valley of a river join the master stream accordingly. But in the glaciated area tributary valleys joining the major glacial trough are quite discordant. The floor of the glacial tributary valleys hang above the floor of the main valley. As glacial erosion is more intensive in the trunk valley, therefore, the trunk valley is wider and deeper than the tributary valley. Once the glacier of the mainstream disappears, the valleys are occupied by stream and the hanging valley turns into waterfall. The Yosemite Valley in the Sierra Nevada of California is an ideal example of a glaciated valley accompanied by hanging valleys, waterfalls and cascades, glaciated spurs etc. Hanging may be produced by tilting of a region which steeper the gradient and accelerate the down-cutting of a main valley without any corresponding effect upon the tributaries. Faulting may also produces hanging valleys. It is not much difficult to distinguish hanging valleys of non-glacial origin from those produced by glaciations.

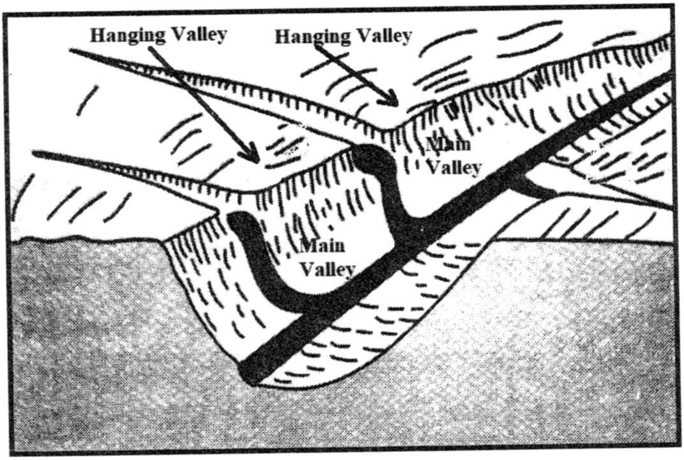

Source: Khullar, D.R.: *Physical Geography* (2014), Kalyani Publishers, Ludhiana, p. 460.

Fig. 84: Hanging Valley.

Fjords

Fjords are unique feature formed by glacial erosion. Fjords are spectacularly existed on the coast of Norway, Greenland, Iceland, southern Alaska, Antarctica etc. where they are mostly buried under ice sheets. Fjords are glacial trough eroded by ice below sea level. They are more common on west facing than east facing coasts and mostly occur in higher latitudes. The depth of water in a fjord is typically much less than seaward or headward part. Thus, a fjord is an elongated basin would become a lake, if sea level were lowered below the top of its threshold. The fjords formed in the coasts are backed by rugged and dissected highlands, which were subjected to sever pliestocene, glaciation. Sea with its long arm stretching inland between steep and rocky mountainous part accommodate the formation of deep and narrow fjords and usually contains hanging tributary valleys. The deepest fjord with a depth of 1,288 m. is located in Chile which is known as Messie channel. Fjords usually possess a submerged rock basin with the deepest part towards the sea-coast. Thus, the floor of the fjord becomes similar to a normal rock basin with threshold at its seaward projected part. There are different views regarding the origin of fjord. It is suggested by some that fjords were essentially tectonic origin. As a result of fracturing under crustal tension, both fluvial and glacial action evacuated the debris and brought slight modification in the form. In some cases, it is true that some fjords show a close

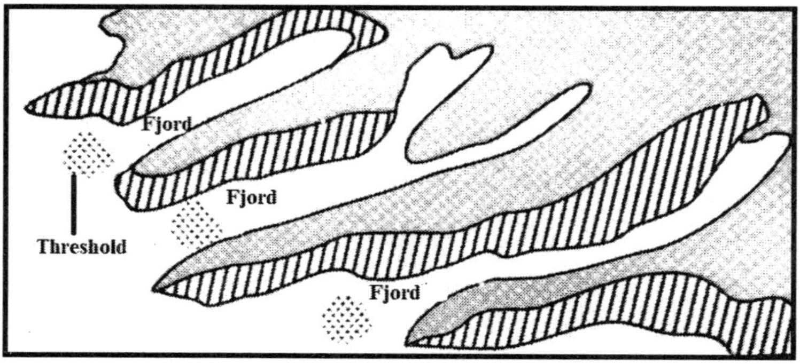

Source: Khullar, D.R.: *Physical Geography* (2014), Kalyani Publishers, Ludhiana, p. 463.

Fig. 85: Fjords.

relationship with the line of structural weakness and some other belief that fractures are not mandatory for the formation of fjords as fjords projected island through normal glacial troughs.

In the same class with fjords are fiards and trough lakes. Fiards are comparatively shorter and shallower than fjords. They are lacking of high relief and developed in the U-shaped cross-section of the fiards. Norway has a large number of fjords white Sweden has as many as fiards. Trough lakes are similar to fjords which are elongated basin with great depth of water and lies above sea level.

Depositional Landform

The debris generated by glacial erosion are not sorted and are characteristically heterogeneous as in content of materials which lacks stratification. These glacial deposits are called **moraine** or **till**. Glacial deposits are distinguished as ground moraine, lateral moraine and end moraine. A blanket of moraine forms beneath a glacier is called **ground moraine**, a thickened rim of moraine at the terminus of glacier is called **end moraine** and moraines found along the sides of a valley glacier is known as **lateral moraine**. Slight oscillations of an ice front as it recedes may develop an irregular belt of knobs and basins, which is described as knob and basin topography. This type of end moraine is usually formed by ice caps rather than by ice streams. Lateral moraines are commonly formed along the sides of an ice stream. Debris are contributed from the valley sides of glaciers by weathering, avalanches, snow slides and other mass movements. **Medial moraine** are produced by the joining of two lateral moraines. Medial moraines are simply a features of glacial surface and not a landform, since it will not persist long after a stream take over the glacial trough. Lateral moraines are most impressive features found along glacial trough. Sometimes, small lakes are found perched above the floor of a glacial trough. These are normally formed between the lateral moraine and the trough wall of a glacial trough.

On the other hand, ground moraine is not abundantly associated with the ice streams. Ground moraine merely concentrated in the ice caps of a glacial surface. Since ice streams are particularly effective erosional agents, therefore most of the debris concentrated in the bed rock is abraded by the flow of ice stream. Ground moraines are occasionally deposited in the valley floor and therefore susceptible for erosion by post-glacial streams flow. Most of the ground moraines are deposited in a thin layer over the valley floor of a glacial stream.

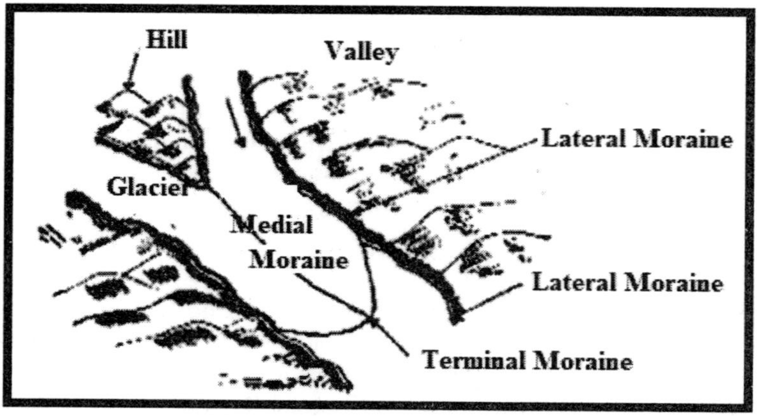

Source: Mitra, Kalpana, Sen, Jyotirmoy & Sengupta Purabi: *Physical Geography* (2014) Kalyani Publisher, Ludhiana, p. 404.

Fig. 86: Moraine.

Fluvio-Glacial Landform

The features produced by the combined action of both glaciers and melt water are called fluvio-glacial landforms. The debris generated and transported by the glacier comes to a total halt at one point and the action of melt-water becomes effective in the transportation and depositional works. The melt-water released by the ice has high ability to perform erosion, transportation and deposition. The landform produced by this process is quite different from landform produced either by glacial and flucial action. The fluvio-glacial deposits retain some of the characteristics of glacial deposits but contain a certain degree of assortment and stratification on account of the action of water. The longer the distance, the debris are transported by the streams before being deposited, the greater the degree of assortment and stratification. The most common fluvio-glacial features are esker, kame, outwash fans, deltas and others.

Fluvio-glacial deposits are usually of two types—pre-glacial deposits and ice-contact deposits. Pro-glacial deposits result from melt water transport which are deposited at a distance where effect of glacier ice is totally absent. On the other hand, ice contact deposits are the product of melt-water sediment in close association of ice mass. These deposits are laid down beneath the ice or on top of the surface or within the ice mass.

Eskers : Eskers are sinuous ridges composed of assorted and stratified sand and gravel. The term '**esker**' is derived from either the Irish word 'eiseir' or welsh word 'escair' meaning crooked or winding. The course of an esker is almost similar to that of a stream, being at times sinuous and at times straight. Eskers are generally found in areas of continental glaciations and seldom found in areas of glacial trough. The low elongated ridge is generally aligned at right angles to the ice front and practically indicates the direction of ice flow. Eskers are usually flat-topped and steep sided. The cross profile of esker shows variation in form depending upon the environment in which they were deposited. The longer ridges are usually higher and wider than the shorter ridges. Eskers between 50 to 200 m. in length range between 3 to 10 m. height while eskers of one kilometer long range in height between 10 to 100 m. Eskers are normally found in north-eastern United States, eastern Canada, Ireland, Finland and Sweden.

Kame terraces : Alike the eskers, Kame is a feature developed in the ice contact zone which is composed of sands and gravels. Generally, Kames are found in form of mound like structure rather than a ridge. Kame is originally a Scottish term introduced into geologic literature by Jamieson in 1874. Holmes defined Kame as "a mound composed chiefly of sand and gravel where form resulted from original deposition modified by any slumping incident to later melting of glacial ice against or upon which the deposit accumulated. Kames are often after found as isolated mounds and exhibit a totally disorganized form. Kames formation are associated with terminal moraines and sometimes found with ground moraine in a scattered position. Numerous mounds when developed as a closely associated group, they are called **kame complex** or **kame group**. Another features may develop as a result of deposition of fluvio-glacial debris in the depressions within the mounds is called **kame and kettle topography**. Another type of kame is developed along the valley side margin of glacial as a result of deposition from melt-water flowing between the ice mass and the valley wall. This type of feature is known as **kame terraces**. Formation of kame terrace is attached with the valley glacier only. It is totally absent in areas of continental glacier.

Although a concrete definition of kame is difficult but all are agreed that it is a mound or hummock composed usually of poorly assorted water-laid materials. Most of the kame are probably of ice contact forms where materials are laid down in a close contact of an ice surface. Kames may have undergone subsequent modification in form by slumping and melting of the ice. Some of the kame represents circular or semi-circular shapes extending down into a glacier which are called **moulin**. Sometimes, kames are formed as small delta cones or fan deposits in small contents along an ice front. Kames are also found in some end moraine belts which are designated as kame moraines.

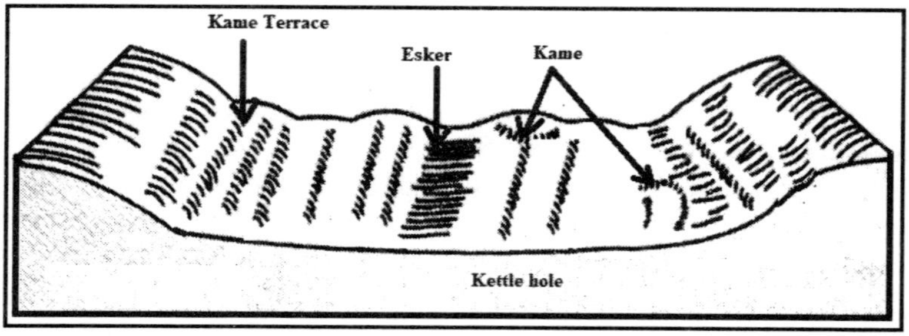

Source: Khullar, D.R.: *Physical Geography* (2014), Kalyani Publishers, Ludhiana, p. 470.

Fig. 87: Kame Terrace, Esker, Kame and Kettle hole.

Till plain Morainic hills

Glacial deposits usually produce two distinctive type of features. The first one is characterized by little local relief and is designated as a **till plain**. These deposits envelop the irregularities present in the bed-rock surfaces and thus becomes thicker over the areas of depression and thinner in areas formed of ridges or plateaus. The surface thus developed exhibits almost plain type topography and is referred as till plain. The second type of feature is characterized by irregular surface, consisting of linear morainic ridges or a complex of ridges, hollows and mounds. Such type of feature is called **hummocky topography**. The hummocky topography is the product of the glacial end moraine while the till plain is formed by the deposition of the ground moraine. Pre-glacial topography is totally marked by the regional deposition of ground moraine and thus buried the pre-glacial topography. Till is unstratified mixture of sand, silt, clay, pebbles and boulders. The materials of the till plain are necessarily derived from bedrocks of nearly areas which differ in character from the bedrock upon which the tills one deposited. Rock fragments deposited on the surface are called **erratics** or **erratic block**. The thickness of the till decrease upward from the glacier's terminus.

Glaciers may deposit till in the form of hills and ridges. As the continental glaciers started retreating from its terminus, the terminal moraines are deposited in places where the glacier halted for some time in the course of retreatment. Back of the outermost end moraine, a series of recessional moraines are found which indicates the successive haltage of retreating ice front. These are found either in the form of linear or irregular ridges interspersed by lake filled depressions. These are called **morainic hills** with an average height of 200 m. from mean sea level. Linear ridges and hills are extensively found in the great European plain.

Drumlins: Some of the till plain possesses streamlined hills on its surface called drumlins. A typical drumlin is half ellipsoid in structure similar to an inverted spoon. Drumlins show a great variation in form, height and length. With an oval outline drumlins are usually several kilometer in length and nearly 60 m. high in average. Drumlins are elongated rounded hills running in the direction of ice movement. Drumlins usually occur in groups called **swarms** or **fields**. Most commonly drumlins are composed of till but some have stratified materials as coating within them and posses a core of bedrock. Drumlins are formed in groups and exhibit a typical feature, which is called **Basket of eggs topography**. These are developed within a short distance from end moraine. Between the rounds of drumlin, the thickness of till is minimum, which as a result give birth to confused drainage, ponds, marshes and waterlogged meadows.

Drumlins normally vary in size, shape, material, and spacing with one common characteristic, which is adhered with their streamlined form. There are two different views regarding the origin of drumlins. The erosional theory explains the origin of drumlins in terms of pre-existing drift cover, means these are formed by two successive glaciations. The objection came from a vital point where question arises regarding the shape of solid rock features. If the drumlins are product of erosion then they should have similarities with the solid rock features which is eventually absent in most of the drumlins. On the contrary, the depositional theory argued that drumlins are formed entirely of till. It is claimed that drumlins formed around a nucleus of frozen till or rock. As few drumlins possess concentric layers of till which, in turn suggest that they have been built up by accretion of layer upon layer. On the other hand, the distribution of drumlins on the lower ground advocates that they develop in areas which are predominantly depositional. I.J. Smalley and D.J. Unwin (1968) forwarded a new theory regarding the formation of drumlin. The content of this theory is related to the dilatancy of the material forming the drumlins. The basic important conditions necessary for the formation of drumlins are (a) the material at the base of ice must have been continuously deformed by shearing, and (b) the material must consist of boulders as like a mixture of clay and water. As dilatancy is a basic property of granular masses, therefore, dilatancy accelerate the pace of drumlin format. They usually develop a layer of boulder-clay representing a glacier from some specific terrain. As flow over this layer is obstructed, part of the layer packs into the obstruction and the rest flows around and moulds it. Expansion results as a result of obstruction imposed on granular mass. The dilatants materials being then more resistant to shear stress, which in turn produces drumlins. Still there are controversies regarding the origin of drumlins but the most common in all cases is the streamlined form of drumlins.

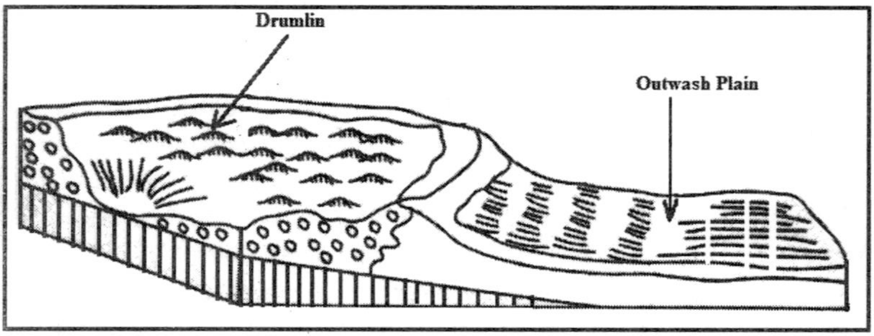

Source: Khullar, D.R.: *Physical Geography* (2014), Kalyani Publishers, Ludhiana, p. 471.

Fig. 88: Drumlin and Outwash Plain.

Glacio-Fluvatile Features

Proglacial form: Proglacial landforms are developed by streams extending beyond an ice front. The major features are outwash fans, deltas and aprons, valley trains and outwash plains. Innumerable small melt-water streams emerge from the glacial trough and built local outwash fans or aprons beyond an ice front or end moraine. Presence of Proglacial Lake immediately beyond the ice margin provides a congenial condition for stream deposition to develop an **outwash plain**. As such, depositional activities by the stream may produce outwash deltas in the margin of the lake. Similarly, where there is an extensive marginal area at the terminating point of melt water streams, the deposition of debris build up an extensive surface which will give birth to a broad outwash plain. **Sandur** is an Icelandic term used as synonyms for outwash plain. Plain sandur or outwash plains have few constraints on their lateral extent and grow as a result of aggradations of a network of braided streams which develops fan shaped features and joining of such depositional plain produce an extensive plain. The debris deposited are of varying size. The coarser materials are deposited at the close of ice margin while the finer particles are carried farther down. Having large load of water and debris, the melt water stream aggraded their valleys and develop braided course and are subject to continuous change of its course.

Most of the outwash plain receive outwash material at the time of advancing stages when the erosional capacity of the glacier is also greater as the melt water streams carry a larger amount of debris. When vigorous flood stream emerge, they promptly erode deep, wide channels. Heavy load of water and eroded materials enforce the rivers to spread out into typical braided pattern and intensify the aggradations of the sandur or outwash plain. Due to the rapid decrease in the discharge of water and debris, the river immediately loses its transporting power and the load it carries are deposited. This process is quite indicative to explain the aggradations that take place or the sandur outwash plain.

Valley trains usually formed between valley walls and usually contain one main channel that spread out in branches to create a network of sub-channels. Valley trains consist mainly of outwash sand and gravel heading usually at an end moraine and extending down valley from it. Glaciers confined between valley walls usually deposited materials in a relatively narrow zone along the valley walls and the resulting feature is known as valley train. Valley trains are marked by terraces above present valley floors. Most of the valleys those have been once glaciated in the past period show some evidence of valley trains. Outwash plains are found extensively in Iceland and occur as fossil forms in New Zealand, Poland, Canada, Germany and United States.

Pitted Outwash Plain: Outwash Plains sometimes are filled with kettle holes which provides a view of hummocky surface. Lumps of ice are left behind man the glacier snout which in turn is enveloped by glacial deposits. At the receding stage that lumps of ice melts and form Kettle holes in the outwash or the sandur. Such type of kettle holes are also found at the beds of large powerful streams as a result of the melting of buried glacial snouts by the fluvio-glacial deposits. The modified outwash plain or the sandur surface with such kettle holes is termed as **kettled Sandur** or **pitted outwash**. The pitted outwash or kettled sandurs are products of proglacial activity. Spectacular development of proglacial deposition along with kettled sandur in the form of great plain is a characteristic feature of Iceland, the land of glacier.

Works of sea waves

Virtually ocean is standing water body only in the sense that the position and margin is almost fixed except minor oscillations produced by the ebb and flow of tides. Movement of water continues all times both at the surface and at depth. Five major processes contribute significantly to marine erosion: **Corrosion, abrasion, attrition, hydraulic action and the shock pressure of breaking waves**. Marine erosion is primarily influenced by several factors. These are:

(a) The kind and durability of rocks along the shore areas.
(b) Structural features of the rocks, particularly its attitude and degree of jointing and fracturing.
(c) Tidal range.
(d) Accessibility of shore to wave attack.
(e) Depth of off-shore water.
(f) Configuration of coastal belts.
(g) Presence or absence of a protective beach.
(h) Availability and size of abrasive tools.
(i) Stability of sea level.

Marine erosion is more pronounced in areas where tools are available but wave impact is intensive upon highly jointed and fractured rocks. Recorded movements of masses weighing many thousands of pounds attest the enormous force exerted by breaking waves. Air lying within the cracks and joints is suddenly compressed and exert tremendous pressure within the rocks. Alternately, recession of the water release the pressure and result in the sudden expansion of air like an explosive. Continuity of the process in turn, exerts great mechanical stress in the rocks, which accelerates the pace of rock disintegration. Sand, gravel pebbles etc. carried by the waves to the shore areas are most effective to erosional processes. Opinion differs in case of the effectiveness of waves in the varying depth of ocean. The term wave base is more often used as the lower limit of wave action. Some scholars agreed that wave action may extend down to depths as great as 600 feet with sufficient force to de-stabilise fine particles on the sea floor.

It is necessary to be familiar with some of the terms, which are used to describe coastal, and shoreline features. The shoreline is a demarcating line where land and water of the sea meet each other. On the other hand, a shore is the zone or area between mean height of low tide and the line of the high tide. This is the inner edge of wave-transported sand. The shore area thus subject to periodic submergence. The water line of the sea periodically migrates over this zone. The coast is the broad extensive zone extending from water line of the sea to the land surface. Sea-cliffs, elevated terraces and lowland areas inside the shore belong to the coastal belt. The part of the shore area, which is covered with unconsolidated debris extending landward from the mean low waterline, is known as **beach**. Due to the wave action, deposition of debris takes place on the **wave cut bench**. When the bench extends seaward into the off-shore zone, this type of feature is called **off-shore bench or beach terrace**. The wave cut bench occasionally may end abruptly with a sharp break in slope towards the sea. Having a gentle seaward gradient, this section of the platform is called **abrasion platform**. The abrasion platform usually contains a thin veneer of fine grained sediments. Again outwards in the sea a zone of sediment deposition may develop which is termed as **continental shore terrace** or **wave built terrace**. Continental shore terrace is joined landward abrasion platform with a thin cover of deposits. The entire part altogether constitutes the continental shelf. Continental shelf is the zone around which there is a marked increase in slope to greater depth. Continental slope on the other hand is the part of the ocean having declivity from the outer edge of the continental shelf. The average gradient of the continental slope is estimated to be $40^{o}17^{/}$ in the upper 6000 feet. The surface of the continental slope is by no means smooth and reliefless. Numerous submarine canyons and furrows produce great relief and in many places, it is marked by submarine hills and ridges. The continental slope represents the zone of juncture between the continental platforms and the ocean basins.

Submarine canyons are the sea valleys, which are to some degree similar to canyons on land. They have V-shaped cross profiles, are sinuous in plan, have tributaries and are essentially cut in the continental slope. Earlier, it was thought that the floors of the ocean basins were largely vast, flat with relatively little relief. Continuous effort to explore the fact of the ocean floor confirmed that the deep sea floors have basin, trenches, troughs, ridges etc. like the continental surface. The term basin is applied to a large depression in the deep sea floor which is somewhat circular, oval or elliptical in plan. A trench, on the other hand, is an elongated narrow depression on the deep sea floor with steep sides, whereas a trough is a long, broad depression with more gently sloping sides. Examples of trenches are : The Mindanao trench (5740 fathoms), Java trench (4074 fathoms), the Aleutian trench (4199 fathoms), Japan trench (5360 fathoms) etc. As such examples of troughs are the Bartlett trough (3958 fathoms), the Weber trough (3451 fathoms) etc. The major topographic features of the deep sea floor are rises, ridges and plateaus. With an width of 600 miles and 1900 miles long, the Hawaiian rise in the Pacific basin is prominent among all rises. Among the mid-oceanic ridges mid-Atlantic ridge, Carlsberg ridge of Indian ocean, Lomonsov ridge of Arctic Ocean, Pacific Antarctic ridge are the prominent examples of oceanic ridges in the world. Submarine elevations of considerable extent with flat tops are called plateaus. Albatross plateaus in the Pacific Ocean, Seychelles plateau of the Indian Ocean, Azores plateau of North Atlantic Ocean etc. are few examples of oceanic plateau.

Seamounts and guyots are two other features developed in the deep sea floor. Seamount is more or less isolated elevation on the ocean floor. The shape of seamounts is somewhat circular or elliptical and possesses vertical relief of at least one kilometer. It is characterized by steep slopes and relatively limited summit area. It is estimated that there may be as many as 10,000 seamounts in the Pacific Ocean alone. On the other hand, guyots are the larger flat-topped seamounts.

This flat topped shape seen to be the result of truncation. How the flat-topped structure came into existence with the Guyots is still a puzzling problem. Once it was concaved that the flat-topped surface was a product of marine planation by Precambrian seas when sea level was much lower than now. Menard (1964) has suggested that more or less flat tops to volcanoes could be produced by erosional truncation near sea level. He favoured the view that Guyots are closely associated with atolls in origin. In the case of an atoll, the volcano reached the sea level and became coated with coral growth and then subsided to form an atoll. Those volcanoes became Guyots, rose to sea level, and were truncated to produce a flat-topped Surface in the Guyots.

Landform Resulting from Marine Erosion

Sea cliffs and wave cut beaches or platforms are two major landform features produced by marine erosion. Along with these two, some other minor features are also produced by marine erosion. Although these are considered as smaller but they are characteristically striking and deserve consideration.

Sea-Cliffs: In several areas, the seaward limit of a coast is marked by a scarp produced by the action sea wave's erosion. This type of feature in general is termed as **sea-cliffs**. The formation of sea-cliff started with the work of erosion on a sloping landmass and cut a small nick or notch. The base of the slope is cut by the waves at the time of high tides and the debris thus produced is transported towards the sea. With increasing erosion, the height of the cliff increases and as such, the width of the wave-cut platform also goes on increasing. The steepness of the cliff increases as the process of marine erosion continues and also develops a spectacular notch at the base of the cliff. As the notch becomes bigger and bigger, the overhanging part of cliff lost its base and become unstable. As the rocks of the cliff started falling down as a result of its foundationless, the cliff continually recedes towards the continental part. Sea cliffs vary greatly depending upon the kind, structure and attitude of the rocks in which they are being cut. Joints and bedding planes are chief determiner of the cliff formation process. Most of the rocks break easily along the joints and bedding planes. Rock strata inclined towards the sea possesses joints, which are inclined towards land. As a result of such structure condition, sea waves easily erode the rocks in size of big blocks. These blocks fall down on the beach and give rise to the cliff in a series of overhanging steps. But under reverse arrangement of rock strata and joints the erosion at the base affects the cliff slowly and the cliff found to be sloping down towards the sea and blocks of rocks easily slide down along the bedding planes and the cliff becomes smooth. Under such condition, cliffs cannot be eroded easily by the sea waves. On the contrary, where the cliffs are composed of hard and resistant rock and absence from having joints and bedding planes the cutting and recession of the cliffs are very slow. Most of notch experience rapid slumping and landslide when the rocks are composed of weak shale or glacial till and thus recession of the cliffs become very fast.

The coastline receding towards the land as a result of wave action is called retrograding coastline. This process is called the **retrogradation of sea cliff**. The process of coastline recession is related to the nature of rocks. Recession is faster where coastline rocks are soft and contains joints and bedding plane in large number. The hard rock coastal areas found projecting into the sea is termed as headlands or buttresses. Most of the semi-circular or semi-oval cover, bays or bights attached with the coastal belt are usually found in parts of the coastal area where rocks are characteristically soft and less resistant to marine erosion. However, effect and intensity of marine erosion greatly differs from region to region and display a close relation with the volume and intensity of sea waves, rock stratigraphy of coastal region, climate condition, topographic characteristics of the coast etc.

Due to wave action in the coastal belt, a number of small but distinctive erosional features are developed. Sea caves are produced in the coast where the coastal rocks are composed of joints or weak dykes. Both mechanical abrasion and chemical erosion takes part in the formation of **sea caves**. Caves are enlarged slowly and the interior of the cave extended a little beyond the reach of waves. As the waves enters into the caves there produce a tremendous pressure by the compression of air inside the cave. As the wave returns, the mouth of the cave become free and compressed air can expand with high intensity. Continuation of this process accelerates the pace of rock breaking. As a result, the size of the cave is enlarged. Sometimes, cracks are developed along the joints of the roof of the caves through which the compressed air inside the cave escape with a hissing sound. These vertical air holes are called **blow-holes**.

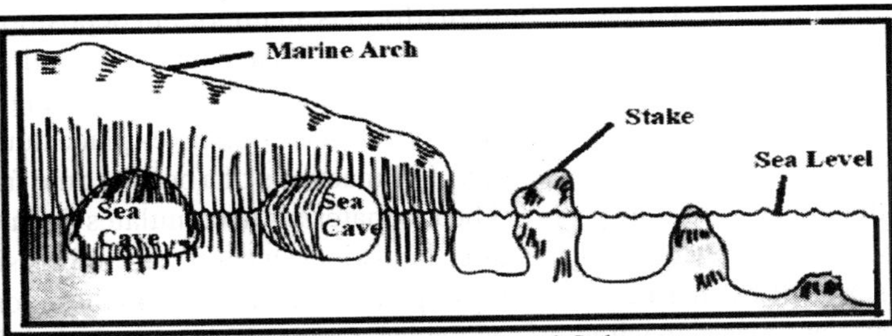

Fig. 89: Marine Arch and Stake.

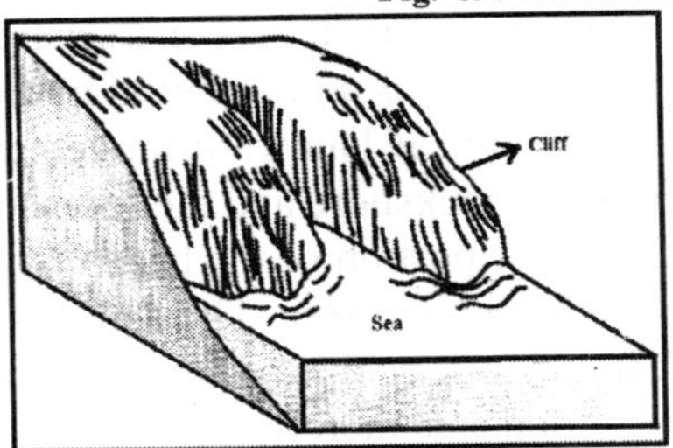

Fig. 90: Cliff.

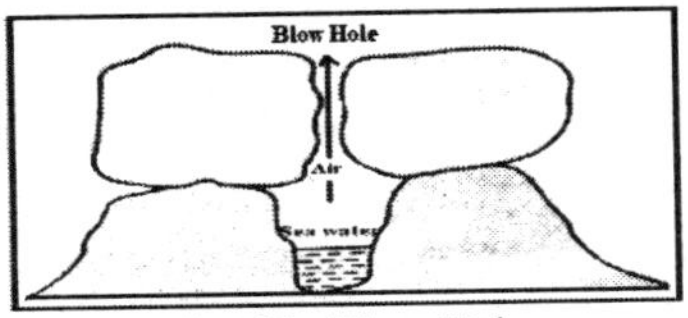

Fig. 91: Blow Hole.

Source: Khullar, D.R.: *Physical Geography* (2014), Kalyani Publishers, Ludhiana, p. 476.

Caves may develop or either side in attachment on the coastal rocks which stands like an entry quite projecting its arms upside having joined by a rocks bar or arch. Such a feature is called **sea arches**. When the arches joining the two-rock pillar fall down or the projecting portion of the headlands is separated from the cliff, isolated rock pillars or vertical pieces of rocks may be found standing on the seacoast. This type of feature is named as **stakes** or **sea stakes**. Arches, stakes, blowholes etc. usually are rare in nature and these are only developed in areas of high wave energy coasts under congenial lithological condition to favour the formation of such features. Sea arches, stakes are found in the Normandy coast of France, Capri Island of Italy and in parts of Sweden and Norway. Caves, arches, buttresses or stakes are considered as secondary features developed as a result of marine erosion.

Landforms Resulting from Marine Deposition

Beaches: Beaches are the wave cut bench along the coastal belt formed by the accumulation of rock debris and are temporary in its character. The beach is a zone of periodic submergence as it is covered by water during high tide. This wave built platform is an important depositional feature in the coast areas of sea. Movement of beach material continues on all times where reduction in beach thickness takes place during the period of extreme storm waves and beach accretion takes place during the period of relatively quite water. The materials of a beach come from both land and sea. Large part of beach materials accumulates from the land, being contributed by streams, landslides, weathering and erosion of sea cliffs by marine waves and slope wash. In addition to these some debris are also deposited by sea waves. Grant (1943) concluded that the oscillatory waves of the sea could transport materials on a sea floor in the direction of their propagation, as there is a differential in their forward and backward velocities in favour of landward movement. The beach is usually formed on both the backshore and the foreshore zone of the coast. Backshore is the upper portion of the beach while the lower most seaward part is called **foreshore area**. The deposits of the beach consist of rounded rock fragments, pebbles, shingles, sands or mixture of all these. The nature of deposits merely depends on the source of the debris and the character of waves.

The beach may be several kilometers long but where the coast is broken into headlands and bays, crescent shaped beaches are formed in the relatively calm and quite waters of the bays. Extensive sandy beaches are found between the mouth of the Mahanadi and Godavari rivers of India. A long and widely stretch beach is formed along the coast where large amount of sands are received from the rivers and the coastal rocks composed of sandstone. The eastern coast of India virtually is the product of depositional activities of the rivers namely Mahanadi, Godavari, Krishna, Cauvery etc.

Bars: The term bar, in generic sense is the various types of submerged or emergent embankments of sand and gravel built on the sea floor by waves and currents. They are usually emergent at the time of low tide. Depending on the form and position, they are known by specific name. Spit is such type of bar which is defined as—**"a ridge or embankment of sediment attached to the land at one end and terminating in open water at the other"** (Evans, 1942). In general, the axis of a spit is straight and almost parallel to the coast except in areas where currents are deflected landward or strong tides exist. Under the condition of strong tidal movement and landward drift of waves, spits take the shape of **re-curved spit** or **hook**.

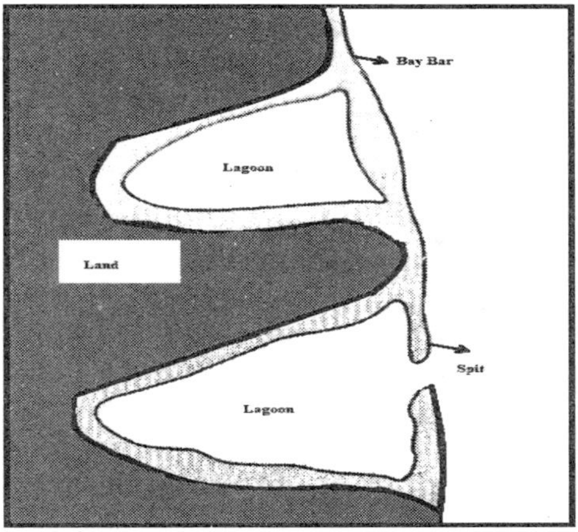

Fig. 92: Bay bar, Spit.

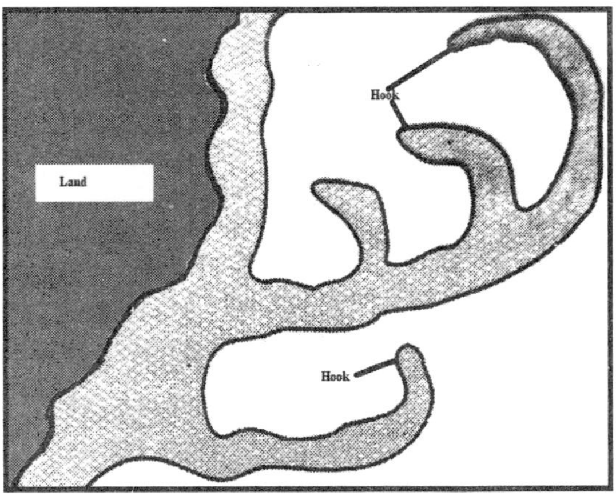

Fig. 93: Lagoon, Hook.

Source: Khullar, D.R.: *Physical Geography* (2014), Kalyani Publishers, Ludhiana, p. 482.

Series of stages of hook development may produce compound hooks. A winged headland is produced when the two ends of a spit is attached with the headland. On the other hand, the offshore convergence of two spits or re-curving of a simple or compound spit without having any attachment to the shore at both ends give birth to a cuspate spit. Successive seaward growth of carpet spit which in turn cause pro-grading of the shoreline and expands seaward produce a feature called cuspate foreland. Florida, Cape Henry, Cape Kennedy etc. are few examples of such features.

Spits are virtually a product of movement and deposition of materials by long shore current or drift which Johnson (1919) designates as wave currents. This type of currents produced by waves meeting a shore obliquely. Movement of materials in this manner is termed as **shore drift** by Johnson. Ocean currents, which are a part of the planetary circulation of the ocean or eddies in it and tidal currents may develop long shore drift. An embankment will be built gradually in the direction on which the current moves. Changes of current directions may develop re-curved spits and other curved shaped features. Tidal currents have no role in the formation of bar as the bars developed in a direction opposite to the tidal currents. According to Lewis (1931, 32) spits are built in a different way where long shore drifts may supply materials to build a spit. Its construction is a result of spasmodic pro-gradation by obliquely impinging waves during major storms. Accordingly, major storm waves that contribute to the permanent part of a bar and the bar extend itself in a direction at right angles to the line of dominant approach by the waves. It was explained that with waves of high frequency, there is a weak development of swash and backwashing of materials from the beach is a precondition and results in the removal of most of the materials from the shore. But low frequency waves developed swash and results in the movement of shingle onto the beach. Thus, storm waves with their high frequency and short wavelengths determine the configuration of spit.

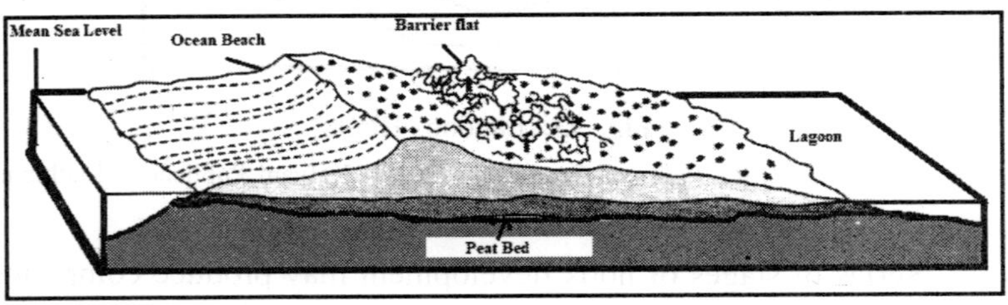

Source: Khullar, D.R.: *Physical Geography* (2014), Kalyani Publishers, Ludhiana, p. 484.

Fig. 94: Features of Barrier Island.

Barrier beaches and barrier chains: More often, the offshore bar is termed as **barrier beach** or **Barrier Island**. Slightly submerged sand ridges are virtually inappropriate to designate as bar. Barrier beaches consist of elongated sand ridges and are parallel to the shore. They usually remain slightly above the water level of high tide. They are separated from the coast by a lagoon. A series of barrier beach extending for or considerable distance is termed as **barrier chain**. In general, the materials in barriers are derived largely from the sea bottom by wave attack. Eastern coast of U.S.A. is a good example of barrier beach. A long and continuous barrier also exists on the east coast of India extending from the Krishna delta in the north to the point Calimere in the south. Continuous tidal lagoons also exist along with the barrier. The barrier varies in width ranging from 3 to 7 kilometer and the sandy deposits rise about 10 meters above sea level.

References

Ahmed. E, (1985). *Geomorphology*, Kalyani Publisher, New Delhi.

Bloom, Arthur, L. (1979). *Geomorphology: A Systematic Analysis of Late Cenozoic Landforms*, Prentice-Hall of India Private Limited, New Delhi.

Dayal, P. (1990). *A Textbook of Geomorphology*, Sukla Book Depot.

Gautam, A. (2006): *Geomorphology*, Sharda Pustak Bhawan, Allahabad.

Hartshorne, R. (1959). *Perspectives on Nature of Geography*, Scientific Publishers.

Husain, Majid (1990). *Fundamentals of Physical Geography,* Rawat Publication.

Khullar, D.R. (2014). *Physical Geography*, Kalyani Publishers, New Delhi.

Mili, Madhurjya, Goswami, B.N., Bora Rana (2019): *Bhu-Akritik Bigyan*, Concept Publishing Company Pvt. Ltd, New Delhi.

NCERT Text Book for class XI: *Fundamentals of Physical Geography*.

Savindra Singh (1998). *Geomorphology*, Prayag Pustak Bhawan.

Sharma, H.S. (ed) (1986). *Geomorphology—Earth Surface Processes and Forms*, Tata McGraw Hill, New Delhi.

Strahler, A.N. (1969). *Physical Geography*, 3rd Edition, Wiley International.

Thronbury, W.D. (1968). *Principles of Geomorphology*, John Wiley & Sons.

Wooldridge S.W. & Margan, R.S. (1959). *An Outline of Geomorphology*, Orient Longman.

Index